BIOMEDICAL NANOTECHNOLOGY

BIOMEDICAL
NANOTECHNOLOGY

Edited by
Neelina H. Malsch

Taylor & Francis
Taylor & Francis Group

Boca Raton   London   New York   Singapore

CRC PRESS, a Taylor & Francis title, part of the Taylor and Francis Group.

Published in 2005 by
CRC Press
Taylor & Francis Group
6000 Broken Sound Parkway NW, Suite 300
Boca Raton, FL 33487-2742

International Standard Book Number-10: 0-8247-2579-4 (Hardcover)
International Standard Book Number-13: 978-0-8247-2579-2 (Hardcover)
Library of Congress Card Number 2005045702

Library of Congress Cataloging-in-Publication Data

Biomedical nanotechnology / edited by Neelina H. Malsch.
    p. cm.
  Includes index.
  ISBN 0-8247-2579-4
  1. Nanotechnology. 2. Medical technology. 3. Biomedical engineering. I. Malsch, Neelina H.

R857.N34B557 2005
610'.28--dc22   2005045702

Preface

In this book, we present the state of the art of nanotechnology research intended for applications in biomedical technologies in three subfields: nanodrugs and drug delivery inside the body; prostheses and implants; and diagnostics and screening technologies for laboratory use. For each of these three subfields, we explore the relevant developments in research.

Nanoparticles such as nanotubes and quantum dots are increasingly applied as drug delivery vehicles. Applications may include gene therapy, cancer treatments, and treatments for HIV and other diseases for which no cures presently exist. Implanted drug delivery or monitoring devices can also include nanostructured materials. Prostheses and implants include nanostructured materials. For example, hip replacements can be made to fit better into the body if coated with nanostructured materials. Nerve tissue can be made to grow along small silicon structures, and this may help paralyzed patients. Nanotechnologies may also contribute to electronic eyes and ears. The research on implants and prostheses focuses on two main directions: (1) biological nanostructures that put biological molecules and tissues in a strait jacket to grow into new structures and (2) biomimetic nanotechnology that starts with physical and chemical structures and aims for a completely new material.

Diagnostics and screening technologies include cantilever biochemical sensors, different types of scanning probe microscopes, lab-on-a-chip techniques, and biosensors. Nanoscience and nanotechnology focus on connecting living materials and electronics as well as on imaging and manipulating individual molecules.

We place these developments in social and economic contexts to assess the likelihood of uptake of these technologies and their relevance to the world's most pressing health needs. Do real needs and markets exist for these devices? We also include a chapter exploring potential risks. The developments in the life science technologies involving GMOs, cloning, and stem cell research have shown that unexpected public concern may slow acceptance of new technologies. For nanotechnology, the public debate is just emerging. Researchers, government officials, and industrialists are actively attempting to assess the risks and redirect research toward the technologies consumers want and away from what the public will not accept.

The scope of this book includes scientific and technological details along with detailed discussions of social and economic contexts. The intended audience includes researchers active in nanoscience and technology in industry and academia, medical professionals, government officials responsible for research, innovation, health care, and biodefense, industrialists in pharmaceutical and biomedical technology, non-governmental organizations interested in environmental, health care, or peace issues, students, and interested lay persons. We assume readers have academic training, but no expertise in nanotechnology.

Contributors

Philip Antón
Rand Corporation
Santa Monica, California

Gabrielle Bloom
Rand Corporation
Santa Monica, California

Ian J. Bruce
Department of Biosciences
University of Kent
Canterbury, Kent, United Kingdom

Aránzazu del Campo
Department of Biosciences
Max-Planck Institut für
  Metallforschung
Stuttgart, Germany

Brian Jackson
Rand Corporation
Santa Monica, California

John A. Jansen
University Medical Center Nijmegen
College of Dental Science
Nijmegen, The Netherlands

Ineke Malsch
Malsch TechnoValuation
Utrecht, The Netherlands

Mark Morrison
Institute of Nanotechnology
Stirling, Scotland
United Kingdom

Mihail C. Roco
National Science Foundation
Chair, U.S. Nanoscale Science,
  Engineering and Technology (NSET)
Washington, D.C.

Emmanuelle Schuler
Rice University
Houston, Texas

Calvin Shipbaugh
Rand Corporation
Santa Monica, California

Richard Silberglitt
Rand Corporation
Santa Monica, California

Jeroen J.J.P. van den Beucken
University Medical Center Nijmegen
College of Dental Science
Nijmegen, The Netherlands

X. Frank Walboomers
University Medical Center Nijmegen
College of Dental Science
Nijmegen, The Netherlands

Kenji Yamamoto, M.D., Ph.D.
Department of Medical Ecology and
  Informatics
Research Institute of the International
  Medical Center
Tokyo, Japan

Contents

Introduction

Converging Technologies: Nanotechnology and Biomedicine

Mihail C. Roco

Recent research on biosystems at the nanoscale has created one of the most dynamic interdisciplinary research and application domains for human discovery and innovation (Figure I.1).* This domain includes better understanding and treatment of living and thinking systems, revolutionary biotechnology processes, synthesis of new drugs and their targeted delivery, regenerative medicine, neuromorphic engineering, and biocompatible materials for sustainable environment. Nanobiosystems and biomedical research are priorities in the United States, the European Union, the United Kingdom, Australia, Japan, Switzerland, China, and other countries and regional organizations.

With proper attention to ethical issues and societal needs, these converging technologies could yield tremendous improvements in human capabilities, societal outcomes, and the quality of life. The worldwide emergence of nanoscale science

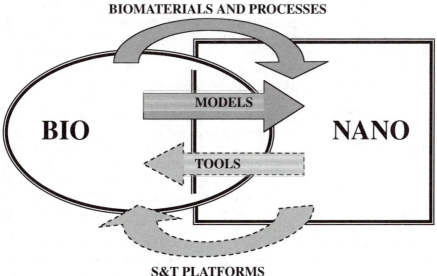

BIOMATERIALS AND PROCESSES

BIO

MODELS

TOOLS

NANO

S&T PLATFORMS

Figure I.1   Interactions of biology and nanotechnology.

---

* The views expressed in this chapter are those of the author and not necessarily those of the U.S. National Science and Technology Council or the National Science Foundation.

and engineering was marked by the announcement of the U.S. National Nanotechnology Initiative (NNI) in January 2000. Its relevance to biomedicine is expected to increase rapidly in the future. The contributions made in this volume are outlined in the context of research directions for the field.

NANOTECHNOLOGY AND NANOBIOMEDICINE

Nanotechnology is the ability to measure, design, and manipulate at the atomic, molecular and supramolecular levels on a scale of about 1 to 100 nm in an effort to understand, create, and use material structures, devices, and systems with fundamentally new properties and functions attributable to their small structures.[1] All biological and man-made systems have their first levels of organization at the nanoscale (nanocrystals, nanotubes, and nanobiomotors), where their fundamental properties and functions are defined. The goal in nanotechnology may be described as the ability to assemble molecules into useful objects hierarchically integrated along several length scales and then, after use, disassemble objects into molecules. Nature already accomplishes this in living systems and in the environment.

Rearranging matter on the nanoscale using "weak" molecular interactions such as van der Waals forces, H bonds, electrostatic dipoles, fluidics, and various surface forces requires low energy consumption and allows for reversible and other subsequent changes. Such changes of usually "soft" nanostructures in a limited temperature range are essential for bioprocesses to take place. Research on "dry" nanostructures is now seeking systematic approaches to engineering human-made objects at nanoscale and integrating nanoscale structures into large-scale structures as nature does. While the specific approaches may be different from the slow evolutions of living systems in aqueous media, many concepts such as self-assembling, templating, interaction on surfaces of various shapes, self-repairing, and integration on multiple length scales can be used as sources of inspiration.

Nanobiomedicine is a field that applies nanoscale principles and techniques to understanding and transforming inert materials and biosystems (nonliving, living or thinking) for medical purposes such as drug synthesis, brain understanding, body part replacement, visualization, and tools for medical interventions. Integration of nanotechnology with biomedicine and biology, and with information technology and cognitive science is expected to accelerate in the next decade.[2] Convergence of nanoscale science with modern biology and medicine is a trend that should be reflected in science policy decisions.[3]

Nanobiosystem science and engineering is one of the most challenging and fastest growing components of nanotechnology. It is essential for better understanding of living systems and for developing new tools for medicine and solutions for health care (such as synthesis of new drugs and their targeted delivery, regenerative medicine, and neuromorphic engineering). One important challenge is understanding the processes inside cells and neural systems. Nanobiosystems are sources of inspiration and provide models for man-made nanosystems. Research may lead to better biocompatible materials and nanobiomaterials for industrial applications. The confluence of biology and nanoscience will contribute to unifying concepts of science, engineering, technology, medicine, and agriculture.

TOWARD MOLECULAR MEDICINE

Nanotechnology provides investigation tools and technology platforms for biomedicine. Examples include working in the subcellular environment, investigating and transforming nanobiosystems (for example, the nervous system) rather than individual nanocomponents, and developing new nanobiosensor platforms. Investigative methods of nanotechnology have made inroads in uncovering fundamental biological processes, including self-assembling, subcellular processes, and system biology (for example, the biology of the neural system).

Key advancements have been made in measurements at the molecular and subcellular levels and in understanding the cell as a highly organized molecular mechanism based on its abilities of information utilization, self-organization, self-repair, and self-replication.[4] Single molecule measurements are shedding light on the dynamic and mechanistic properties of molecular biomachines, both *in vivo* and *in vitro*, allowing direct investigation of molecular motors, enzyme reactions, protein dynamics, DNA transcription, and cell signaling. Chemical composition has been measured within a cell *in vivo*.

Another trend is the transition from understanding and control of a single nanostructure to nanosystems. We are beginning to understand the interactions of subcellular components and the molecular origins of diseases. This has implications in the areas of medical diagnostics, treatments, and human tissue replacements. Spatial and temporal interactions of cells including intracellular forces have been measured. Atomic force microscopy has been used to measure intermolecular binding strength of a pair of molecules in a physiological solution, providing quantitative evidence of their cohesive function.[5] Flows and forces around cells have been quantitatively determined, and mechanics of biomolecules are better understood.[6] It is accepted that cell architecture and macro behavior are determined by small-scale intercellular interactions.

Other trends include the ability to detect molecular phenomena and build sensors and systems of sensors that have high degrees of accuracy and cover large domains. Fluorescent semiconductor nanoparticles or quantum dots can be used in imaging as markers for biological processes because they photobleach much more slowly than dye molecules and their emission wave lengths can be finely tuned. Key challenges are the encapsulation of nanoparticles with biocompatible layers and avoiding nonspecific adsorption. Nanoscience investigative tools help us understand self-organization, supramolecular chemistry and assembly dynamics, and self-assembly of nanoscopic, mesoscopic, and even macroscopic components of living systems.[7]

Emerging areas include developing realistic molecular modeling for "soft" matter,[8] obtaining nonensemble-averaged information at the nanoscale, understanding energy supply and conversion to cells (photons and lasers), and regeneration mechanisms. Because the first level of organization of all living systems is at the nanoscale, it is expected that nanotechnology will affect almost all branches of medicine. This volume discusses important contributions in key areas. In Chapter 1, Morrison and Malsch discuss worldwide trends in biomedical nanotechnology programs. They cover the efforts of governments, academia, research organizations, and other entities related to biomedical nanotechnology.

DRUG SYNTHESIS AND DELIVERY

Yamamoto (Chapter 2) discusses the new contributions of nanotechnology in comparison to existing methods to release, target, and control drug delivery inside the human body. Self-assembly and self-organization of matter offer new pathways for achieving desired properties and functions. Exploiting nanoparticle sizes and nanosized gaps between structures represent other ways of obtaining new properties and physical access inside tissues and cells. Quantum dots are used for visualization in drug delivery because of their fluorescence and ability to trace very small biological structures. The secondary effects of the new techniques include raising safety concerns such as toxicity that must be addressed before the techniques are used in medical practice.

IMPLANTS AND PROSTHESES

Van den Beucken et al. (Chapter 3) demonstrates how nanotechnology approaches for biocompatible implants and prostheses become more relevant as life expectancy increases. The main challenges are the synthesis of biocompatible materials, understanding and eventually controlling the biological processes that occur upon implantation of natural materials and synthetic devices, and identifying future applications of biomedical nanotechnology to address various health issues. The use of currently available nanofabrication methods for implants and understanding cell behavior when brought in contact with nanostructured materials are also described.

DIAGNOSTICS AND SCREENING

Del Campo and Bruce (Chapter 4) review the potential of nanotechnology for high throughput screening. The complexity and diversity of biomolecules and the range of external agents affecting biomolecules underline the importance of this capability. The current approaches and future trends are outlined for various groups of diseases, tissue lapping, and therapeutics. The most successful methods are based on flat surface and fiberoptic microarrays, microfluidics, and quantum dots.

Nanoscale sensors and their integration into biological and chemical detection devices for defense purposes are reviewed by Shipbaugh et al. (Chapter 5). Typical threats and solutions for measuring, networking, and transmitting information are presented. Airborne and contact exposures can be evaluated using nanoscale principles of operation for sensing. Key challenges for future research for biological and chemical detection are outlined.[8]

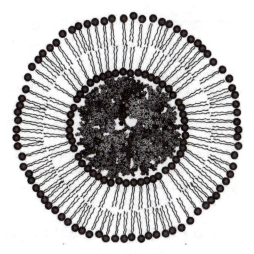

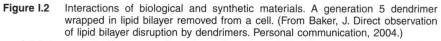

Figure I.2   Interactions of biological and synthetic materials. A generation 5 dendrimer wrapped in lipid bilayer removed from a cell. (From Baker, J. Direct observation of lipid bilayer disruption by dendrimers. Personal communication, 2004.)

One example of the complexity of the scientific issues identified at the interface between synthetic and biological materials and systems is the study of toxicity caused by dendrimers.[9] Generation 5 dendrimers of particular diameters and electrically and positively charged can actually rip lipid bilayers from cells to form micellar-like structures (Figure I.2), leading to cytotoxicity. The health concerns caused by nanotechnology products must receive full consideration from the private sector and government organizations because of the specific properties and types of complex interactions at the nanoscale.

NANOTECHNOLOGY PLATFORMS FOR BIOMEDICINE

Nanotechnology offers new solutions for the transformation of biosystems and provides a broad technological platform for applications in industry; such applications include bioprocessing, molecular medicine (detection and treatment of illnesses, body part replacement, regenerative medicine, nanoscale surgery, synthesis and targeted delivery of drugs), environmental improvement (mitigation of pollution and ecotoxicology), improving food and agricultural systems (enhancing agricultural output, new food products, food conservation), and improving human performance (enhancing sensorial capacity, connecting brain and mind, integrating neural systems with nanoelectronics and nanostructured materials).

Nanotechnology will also serve as a technological platform for new developments in biotechnology; for example, biochips, "green" manufacturing (biocompatibility and biocomplexity aspects), sensors for astronauts and soldiers, biofluidics for handling DNA and other molecules, *in vitro* fertilization for livestock, nanofiltration, bioprocessing by design, and traceability of genetically modified foods.

Exploratory areas include understanding, conditioning, and repairing brain and other parts for regaining cognition, pharmaceuticals and plant genomes, synthesis of more effective and biodegradable chemicals for agriculture, implantable detectors, and use of saliva instead of blood for detection of illnesses. Broader issues include economic molecular medicine, sustainable agriculture, conservation of biocomplexity, and enabling emerging technologies. Measurements of biological entities such as neural systems may be possible at the level of developing interneuronal synapse circuits and their 20-nm diameter synoptic vesicles. Other potential breakthroughs that may be targeted by the research community in the next 10 years are the detection and treatment of cancer, treatment of brain illnesses, understanding and addressing chronic illnesses, improving human sensorial capacity, maintaining quality of life throughout the aging process, and enhancing learning capabilities.

FUNDING AND POLICY IMPLICATIONS

With proper attention to ethical issues and societal needs, these converging technologies could allow tremendous improvements in human capabilities, societal outcomes, and the quality of life. Malsch (Chapter 6) examines the potential of nanotechnology to address health care needs and the societal implications of nano-biomedical research and development. The most important avenues of disease treatment and the main issues to be considered by governments, civic organizations, and the public are evaluated. The social, economic, ethical, and legal aspects are integral parts of nanotechnology R&D for biomedical applications.

Schuler (Chapter 7) reviews the potential risks of biomedical nanotechnology and outlines several scenarios for eventual regulation via market forces, extensions of current regulations, accidents, regulatory capture, self-regulation, or technology ban. The chances of success of these scenarios are determined by the way the stakeholders respond to the large-scale production and commercialization expected to begin within the next decade.

The United States initiated a multidisciplinary strategy for development of science and engineering fundamentals through its NNI in 2000. Japan and Europe now have broad programs and plans for the next 4 or 5 years. More than 40 countries have developed programs or focused projects in nanotechnology since 2000. Research on biosystems has received larger support in the United States, the United Kingdom, Germany, Switzerland, and Japan. Other significant investments in nanotechnology research programs with contributions to nanobiosystems have been made by the European Community, Australia, Taiwan, Canada, Finland, Italy, Israel, Singapore, and Sweden. Relatively large programs in nanotechnology but with small biosystems components until 2004 have been developed by South Korea and China. Worldwide government funding has increased to about eight times what it was in 1997, exceeding $3.6 billion in 2004 (see http://www.nsf.gov/nano). Differences among countries can be noted by the research domains they choose, the levels of program integration into various industrial sectors, and the time scales of their R&D targets.

Of the total NNI investment in 2004, about 15% is dedicated to nanobiosystems in two ways. First, the implementation plan of NNI focuses on fundamental research related to nanobiosystems and nanomedicine. Second, the program involves two grand challenges related to health issues and bionanodevices. Additional investments have been made for development of infrastructures at various NSF centers, including the Cornell University Nanotechnology Center and additional nanoscale science and engineering centers at Rice University, the University of Pennsylvania, and Ohio State University.

The NNI was evaluated by the National Research Council and the council published its findings in June 2002. One recommendation was to expand research at the interface of nanoscale technology with biology, biotechnology, and life sciences. Such plans to extend nanobiosystems research are under way at the U.S. Department of Energy (DOE), the National Institutes of Health (NIH), the National Science Foundation (NSF), and the Department of Agriculture (USDA). A NSF–Department of Commerce (DOC) report recommends a focus on improving physical and mental human performance through converging technologies.[2] The NSF, the National Aeronautics & Space Administration (NASA), and the Department of Defense (DOD) have included aspects of converging technologies and improving human performance in their program solicitations. The Defense Advanced Research Projects Agency (DARPA) instituted a program on engineered biomolecular nanodevices and systems. A letter sent to the NIH director by seven US senators in 2003 recommended that the NIH increase funding in nanotechnology. The White House budget request for fiscal 2004 lists "nanobiosystems for medical advances and new products" as a priority within the NNI. Nanobiotechnology RRD is highlighted in the long-term NNI Strategic Plan published in December 2004 (http://www.nano.gov). Public interactions provide feedback for the societal acceptance of nanotechnology, and particularly the aspects related to human dimensions and nanobiotechnology.[10,11]

Nanobiosystems is an area of interest recognized by various international studies on nanotechnology, such as those prepared by Asia-Pacific Economic Council (APEC),[12] the Meridian Institute,[13] and Economic Organization of Developed Countries (OECD).[14] In a survey performed by the United Kingdom Institute of Nanotechnology and by OECD,[14] experts identified the locations of the most sophisticated nanotechnology developments in the medical and pharmaceutical areas in the United States (48%), the United Kingdom (20%), Germany (17%), Switzerland (8%), Sweden (4%), and Japan (3%). The U.S. NNI plans to devote about 15% of its fiscal year 2004 budget to nanobiosystems; Germany will allocate about 10% and France about 8%. The biology route to nanotechnology may be a choice for countries with less developed economies because required research facility investments are lower.

CLOSING REMARKS

Nanoscale and biosystem research areas are merging with information technology and cognitive science, leading to completely new science and technology platforms in genome pharmaceuticals, biosystem-on-a-chip devices, regenerative medicine, neuroscience, and food systems. A key challenge is bringing together biologists and doctors with scientists and engineers interested in the new measurement and fabrication capabilities of nanotechnology. Another key challenge is forecasting and addressing possible unexpected consequences of the revolutionary systems and engineering developments utilized in nanobiosystems. Priority science and technology goals may be envisioned for international collaboration in nanoscale research and education, better comprehension of nature, increasing productivity, sustainable development, and addressing humanity and civilization issues.

The confluence of biology, medicine, and nanotechnology is reflected in government funding programs and science policies. For example, the U.S. NNI plans to increase its contributions to programs dedicated to nanobiosystems beyond the current level of about 15%; similar trends in other countries intended to better recognize nanobiosystems research have also been noted.

Nanoscale assemblies of organic and inorganic matter lead to the formation of cells and other activities of the most complex known systems — the human brain and body. Nanotechnology plays a key role in understanding these processes and the advancement of biological sciences, biotechnology, and medicine. Four chapters in this volume present key issues of molecular medicine, from drug delivery and biocompatible replacement body parts to devices and systems for high throughput diagnostics and biodefense. Three other chapters provide overviews on relevant research and development programs, the social and economic contexts, and potential uncertainties surrounding nanobiomedical developments. This broad perspective is of interest not only to the scientific and medical community, but also to science policy makers, social scientists, economists, and the public.

REFERENCES

1. Roco MC, Williams RS, and Alivisatos P, Eds. *Nanotechnology Research Directions.* Kluwer Academic Publishers, Dordrecht, 2000, chap. 8.
2. Roco MC and Bainbridge WS, Eds. Converging Technologies for Improving Human Performance. National Science Foundation–U.S. Department of Commerce Report, Washington, D.C., 2002.
3. Roco MC. Nanotechnology: convergence with modern biology and medicine. *Curr Opinion Biotechnol* 14: 2003, 337–346.
4. Ishijima A and Yanagida T. Single molecule nanobioscience. *Trends Biochem Sci* 26: 438–444, 2001.
5. Misevic GN. Atomic force microscopy measurements: binding strength between a single pair of molecules in physiological solutions. *Mol Biotechnol* 18: 149–154, 2001.
6. Bao G. Mechanics of biomolecules. *J Mech Physics Solids* 50: 2237–2274, 2002.
7. Whitesides G and Boncheva M. Beyond molecules: self-assembling of mesoscopic and macroscopic components. *Proc Natl Acad Sci USA* 99: 4769–4774, 2002.
8. Nielaba P, Mareschal M, and Ciccotti G, Eds. *Bridging the Time Scales: Molecular Simulations for the Next Decade,* Springer, New York, 2002.
9. Baker J. Direct observation of lipid bilayer disruption by dendrimers, personal communication, 2004.

10. Bainbridge WS. 2002. Public attitudes toward nanotechnology. *J Nanoparticle Res* 4: 461–464, 2002.
11. Cobb MD and Macoubrie J. 2004. Public perceptions about nanotechnology: benefits, risks and trust. *J. Nanoparticle Res* 6: 2004, 395–405.
12. APEC (Asia-Pacific Economic Council). Nanotechnology: the technology for the 21st century, Report, Bangkok, Thailand, August 2001.
13. Meridian Institute. Summary of the International Dialog for Responsible R&D of Nanotechnology. National Science Foundation, Alexandria, VA, 2004.
14. OECD. Nanotechnology R&D programs in the U.S., Japan and the European Union: preliminary review. Working Party on Innovation and Technology Policy, Paris, December 10–11, 2002.

CHAPTER 1

Trends in Biomedical Nanotechnology Programs Worldwide

Mark Morrison and Ineke Malsch

CONTENTS

I. INTRODUCTION

This chapter covers an overview of trends in nanotechnology research programs for biomedical applications in the United States, leading European countries, and Japan. We focus on technologies for applications inside the body, including drug delivery technologies for pharmaceuticals, and new materials and technologies for prostheses and implants. We also include technologies for applications outside the body including diagnostics and high throughput screening of drug compounds. We cover the main application areas in pharmaceuticals and medical devices — areas where governments expect nanotechnology to make important contributions. We also outline the currently operational national and European Union (EU) policies and programs intended to stimulate the development of biomedical nanotechnology in the U.S., Europe, and Japan.

Several applications of nanotechnology are already available in the market. Lipid spheres (liposomes) with diameters of 100 nm are available for carrying anticancer drugs inside the body. Some anti-fungal foot sprays contain nanoscale zinc oxide particles to reduce clogging.

Nanotechnology is producing short-term impacts in the areas of:

Medical diagnostic tools and sensors
Drug delivery
Catalysts (many applications in chemistry and pharmaceuticals)
Alloys (e.g., steel and materials used in prosthetics)
Improved and body-friendly implants
Biosensors and chemical sensors
Bioanalysis tools
Bioseparation technologies
Medical imaging
Filters

Most current applications utilize nanopowder qualities instead of other properties present at the nanoscale. The next stage of applications of nanotechnology will allow products to exhibit more unusual properties as product creation is approached from the bottom up. This is considered a measure of the development of nanotechnology. Long-term product and application perspectives of nanotechnology with high future market potentials include:

Perfect selective sensors for the control of environment, food, and body functions
Pharmaceuticals that have long-term dosable capabilities and can be taken orally
Replacements for human tissues and organs
Economical or reusable diagnostic chips for preventive medical surveys

It is estimated that more than 300 companies in Europe are involved in nano-technology as their primary areas of business, and many more companies, particu-larly larger organizations, are pursuing some activities in the field. Large organiza-tions currently exploring the possibilities of nanotechnology with near-term applications in drug delivery are Biosante, Akzo Nobel, Ciba, Eli Lilly, and Merck.

II. BIOMEDICAL NANOTECHNOLOGY IN THE UNITED STATES

A. National Nanotechnology Initiative

The National Nanotechnology Initiative (NNI) in the United States is built around five funding themes distributed among the agencies currently funding nanoscale science and technology (S&T) research (see Table 1.1). In addition to federal fund-ing, the individual states are also dedicating considerable funds to nanotechnology. Long-term basic nanoscience and engineering research currently focuses on funda-mental understanding and synthesis of nanometer-size building blocks aimed at potential breakthroughs in several areas including medicine and health care, the chemical and pharmaceutical industries, biotechnology and agriculture, and national security. This funding is intended to provide sustained support for individual inves-tigators and small groups performing fundamental research, promote univer-sity–industry–federal laboratory partnerships, and foster interagency collaboration.

The Grand Challenges theme of the initiative includes support for interdiscipli-nary research and education teams including centers and networks that work on key long-term objectives. The Bush administration identified a dozen grand challenges essential for the advancement of nanoscale science and technology. They include the design and manufacture of nanostructured materials that are correct at the atomic and single-molecule levels. These advances are aimed at applications including biological sensors for use in health care and chemical and biological threat detection.

Table 1.1 United States National Nanotechnology Initiative Budget by Agency*

Department or Agency	FY 1999	FY 2000	FY 2001	FY 2002	FY 2003	FY 2004	FY 2005
Dept of Defense	70	70	123	180	322	315	276
Environmental Protection Agency		–	5	5	5	5	5
National Aeronautics and Space Administration	5	5	22	46	36	37	35
National Institutes of Health	21	32	39.6	40.8	78	80	89
National Science Foundation	85	97	150	199	221	254	305
Total	225	270	463.85	604.4	862	961	982

* In millions of dollars.

Many of the challenges are aligned with the missions of the various agencies participating in the NNI. We describe the activities of some of these agencies in the area of biomedical nanotechnology later in this chapter.

Ten centers and networks of excellence have been established, each of which has been granted funding of about $3 million annually for 5 years. Pending a successful interim progress review, each center may be eligible for a one-time 5-year renewal. The centers will play a key role in achieving top NNI priorities (fundamental research, grand challenges, educating future scientists and engineers) in developing and utilizing specific nanoscale research tools and in promoting research partnerships. It is anticipated that the establishment of centers and networks will aid the integration of research and education in nanoscale science and technology across disciplines and various research sectors including universities, federal laboratories, and the private sector. Interdisciplinary research activities of government, university, and industrial performers will create a vertical integration arrangement with expertise ranging from basic research to the development of specific nanotechnology devices and applications.

The NNI also supports the creation of a research infrastructure for metrology, instrumentation, modeling and simulation, and facilities. Work at the nanoscale requires new research tools, for example, new forms of lithography, computational capabilities, and instruments for manipulation. New research centers possessing such instrumentation will be built and made available to researchers from universities, industries, and government laboratories. The ultimate objective is to develop innovations that can be rapidly commercialized by United States industries. According to the Nanoscale Science and Engineering (NSE) Group representatives, if the need for instrumentation and the ability to make the transition from knowledge-driven to product-driven efforts are not addressed satisfactorily, the United States will not remain internationally competitive in this field.

The societal implications of nanotechnology and workforce education and training constitute the fifth theme of the NNI. In concert with the initiative's university-based research activities, this effort is designed to educate and train skilled workers, giving them the interdisciplinary perspective necessary for rapid progress in nanoscale science and technology. Researchers will also examine the potential ethical, legal, social, and workforce implications of nanoscale science and technology.

In fiscal year (FY) 2002, the NNI initiative focused on long-term research investigating the manipulation of matter at the atomic and molecular levels. This research may lead to continued improvements in electronics for information technology; higher performance, lower maintenance materials for manufacturing, defense, transportation, space, and environmental applications; and accelerated biotechnological applications for medicine, health care, and agriculture. New areas of research and development focus initiated in all federal departments and agencies in 2003 included the uses of nanotechnology for chemical–biological–radioactive–explosive (CBRE) detection and protection. The NNI Initiative also focuses on fundamental nanoscale research through investments in investigator-led activities, centers and networks of excellence, and infrastructure. In 2004, the NNI added two biomedical related priorities: (1) nanobiological systems for medical advances and new products, and (2) nanotechnology solutions for detection of and protection from weapons of mass destruction.

B. Federal Agencies

According to the NNI implementation plan, each agency invests in projects that support its own mission and retains control over how it will allocate resources against its NNI proposals based on the availability of funding. Each agency evaluates its own NNI research activities according to Government Performance and Results Act (GPRA) procedures. Most of the funding by government agencies is generally allocated to proposals submitted in response to program announcements and initiatives and selected by a peer review process.

1. National Science Foundation

The National Science Foundation (NSF) has five programmatic focus areas:

1. Fundamental research and education with special emphasis on biosystems at nanoscale level; nanoscale structures, novel phenomena, and quantum control; device and system architecture; nanoscale processes in the environment, and manufacturing processes at nanoscale; multiscale, multiphenomena theory, modeling and simulation at nanoscale.
2. Grand Challenges funding of interdisciplinary activities focusing on major long-term challenges: nanostructured materials by design, nanoscale electronics, optoelectronics and magnetics, nanoscale-based manufacturing, catalysts, chemical manufacturing, environment, and health care.
3. Centers and networks of excellence to provide support for about 15 research and education centers that will constitute a multidisciplinary, multisectorial network for modeling and simulation at nanoscale and nanofabrication experimentation and user facilities; see below.
4. Research infrastructure for instrumentation and facilities for improved measurements, processing and manipulation at nanoscale, and equipment and software for modeling and simulation.
5. Societal and educational implications of science and technology advances for student assistantships, fellowships, and traineeships; curriculum development related to nanoscience and engineering and development of new teaching tools.

The impacts of nanotechnology on society will be analyzed from legal, ethical, social, and economic perspectives. Collaborative activities with the National Aeronautics & Space Administration (NASA) related to nanobiotechnology and nanodevices and with the National Institutes of Health (NIH) in the fields of bioengineering and bionanodevices will be planned. The NSE Group, including representatives from all directorates, will coordinate the NNI activities at the National Science Foundation (NSF). Each directorate will have two representatives in the NSE Group and the chair is the NSF representative. The nanotechnology research centers supported by NSF focus on specific areas of nanoscale science and engineering and participate in collaborations with industries and other institutions.

a. Nanobiotechnology Center at Cornell University

The NSF established the Nanobiotechnology Center (NBTC) at Cornell University as a science and technology facility in 2000. The NBTC applies the tools and processes of nano- and microfabrication to build devices for studying biosystems and learning from biology how to create better micro-nanoscale devices. The center's work involves nanofabricated materials that incorporate cellular components on their own length scales, for example, proteins and DNA, and nanobiotechnology that offers opportunities of biological functionalities provided by evolution and presents challenges at the inorganic–biological interface. The center utilizes nanofabricated research tools to probe biological systems, separate biological components for characterization, and engineer biological components within useful devices.

b. National Nanofabrication Users Network

Created in 1993, the National Nanofabrication Users Network (NNUN) gives researchers access to advanced equipment. Facilities at five major universities comprise the network that supported about 1100 graduate and undergraduate researchers in 2001. Plans are underway to add centers and tie other government facilities into the NNUN. The network currently consists of two hub facilities on the east and west coasts (at Cornell University in Ithaca, New York, and at Stanford University in Palo Alto, California) and three additional centers at Howard University (Washington, D.C.), Pennsylvania State University, and the University of California at Santa Barbara that offer expertise in specific areas.

c. Columbia University

Columbia University includes the Center for Electronic Transport in Molecular Nanostructures. The center works with industry and national laboratories to explain the effects of charges in applications such as electronics, photonics, and medicine. The Columbia center conducts research that will establish the foundations for new paradigms for information processing through the fundamental understanding of charge transport phenomena unique to nanoscale molecular structures. The center's research program addresses electronic transport in molecular nanostructure; it also designs insulators for molecular circuitry and builds molecules that can handle the operational functions of a transistor.

d. Northwestern University

Northwestern University's Center for Integrated Nanopatterning and Detection Technologies is headed by Chad Mirkin. The NSE's Center for Integrated Detection and Patterning Technologies focuses on the development of state-of-the-art nano-patterning and detection devices. The center's innovative nanoscience work is aimed at receptor design, signal transduction, systems integration, and new technology in the areas of biodiagnostics and high throughput screening.

e. Rensselaer Polytechnic University

Richard Siegel is the director of Rensselaer Polytechnic University's Center for Directed Assembly of Nanostructures. The center works with the University of Illinois at Urbana–Champaign and the Los Alamos National Laboratory in New Mexico on materials projects involving composites, drug delivery devices, and sensors. Research projects include investigations of functional nanocomposites that may find use in a variety of structural, electrical, and biomedical applications.

f. Rice University

Rice University is the site of the Center for Biological and Environmental Nanotechnology; the co-directors are Richard Smalley and Vicki Colvin. The center focuses on bioengineering and environmental engineering with emphases on nanoscale biology and chemistry. The center's work encompasses nanomaterials for bioengineering applications, including developing medical therapeutics and diagnostics and environmental science and engineering. It also works on developing nanomaterial solutions to persistent environmental engineering problems.

2. Department of Defense

Nanotechnology continues to be one of the top priority research programs within the U.S. Department of Defense (DOD). The department's investment in nanotechnology is organized to focus on three nanotechnology areas of critical importance including nanobiodevices. The DOD structures its science and technology investments into basic research, applied research, and exploratory development. The latter two focus on transitioning science discovery into innovative technology. Several general technology transfer programs are also available for transition efforts.

In 1999 and 2000, one of the main aspects of nanotechnology related to chemical and biological warfare defense. Particular priorities were novel phenomena, processes, and tools for characterization and manipulation ($19 million) and biochemical sensing ($1 million). Modes of research and development (R&D) support were principally university-based programs for individual investigators and centers, certain programs at DOD laboratories, and infrastructure (equipment, high performance computing). FY 2002 funding was utilized to augment programs in the three NNI R&D Grand Challenges with particular DOD interest focused on bionanosensor devices.

The Defense Advanced Research Projects Agency (DARPA) undertook significant enhancements in nanoscience nanotechnology projects in its investment portfolio in FY 2003. New programs include nanostructures in biology and quantum information S&T. The increase is consistent with the Quadrennial Defense Review recommencing expansion of the S&T budget to 3% of the DOD budget.

The events of September 11, 2001 motivated accelerated concentration on innovative technologies to improve the national security posture relative to chemical, biological, radiological, and explosive substances. DOD will play a major role in this multiagency effort. Its Advisory Group on Electronic Devices (AGED) performed a special technical area review (STAR) of nanoelectronics. Key goals of the review were guidance for the basic science investments in nanoelectronics, opto-electronics, and magnetics and the funding necessary to accelerate the development of information technology devices.

The U.S. Army allocated $10 million in basic research funds for a university-affiliated research center (UARC) designated the Institute for Soldier Nanotechnol-ogies (ISN). The Naval Research Laboratory formed a nanoscience institute to enhance multidisciplinary thinking and critical infrastructure. The mission of the institute is to conduct highly innovative interdisciplinary research at the intersections of the nanometer-sized materials, electronics, and biology domains. The institute is making progress in the high-density nonvolatile memory, biological and chemical sensor, and biological–electronic interface areas.

a. Institute for Soldier Nanotechnologies

Massachusetts Institute of Technology (MIT) has been selected to host the ISN. The purpose of this research center of excellence is to develop unclassified nano-meter-scale S&T solutions for soldiers. The anticipated basic research effort is to be funded between FY 2002 and FY 2006 and amounts to $50 million. An additional $20 million may also be provided in the form of subsequent UARC subcontracts for accelerated transition of concepts into producible technologies by industrial partners participating in research at the ISN. Industry will contribute an additional $40 million in funds and equipment.

The ISN will be staffed by up to 150 people, including 35 MIT professors from 9 departments in the schools of engineering, science, and architecture and planning. In addition to faculty, 80 graduate students, and 20 postdoctoral associates, the ISN will also include specialists from the U.S. Army, DuPont, Raytheon, Massachusetts General Hospital, and Brigham and Women's Hospital. The two hospitals and MIT are also members of the Center for Integration of Medicine and Innovative Tech-nology. The ISN will focus on six key soldier capabilities: (1) threat detection, (2) threat neutralization, (3) concealment, (4) enhanced human performance, (5) real-time automated medical treatment, and (6) reduced logistical footprints. The themes to be addressed by seven research teams are:

1. Energy-absorbing materials
2. Mechanically active materials for devices and exoskeletons
3. Detection and signature management
4. Biomaterials and nanodevices for soldier medical technology
5. Systems for manufacture and processing of materials
6. Modeling and simulation
7. Systems integration

Raytheon, DuPont, and the two hospitals serve as founding industrial partners that will work closely with the ISN and with the Army Natick Soldier Center and Research Laboratory to advance the science of field-ready products.

3. National Aeronautics and Space Administration

A major focus of NASA is advancing and exploiting the zone of convergence of nanotechnology, biotechnology, and information technology related to space exploration. NASA envisions aerospace vehicles and spacecraft made from materials ten times stronger and less than half the weights of current materials. Such equipment will include embedded sensors, actuators, and devices to monitor internal health *in situ* during extended space missions and perform self-repairs of vehicles. Information systems and science systems based on nanoscale electronics will extend beyond the limits of silicon, leading to the capability to conduct complex missions nearly autonomously. Key areas of NASA research and technology development involve high performance aerospace materials including carbon nanotube and high temperature nanoscale composites; ultrahigh density, low power, and space-durable information systems, electronics, and sensor systems; ultrasensitive and robust spacecraft systems; and systems for *in situ* human health care.

NASA's investmens in nanoscience and nanotechnology involve contributions of several laboratories (mainly Ames, Langley, and the Jet Propulsion Laboratory [JPL]) and externally supported research. In 2001, the priorities in nanotechnology included biomedical sensors and medical devices. Major themes and new programs in FY 2002 were:

Manufacturing techniques for single-walled carbon nanotubes for structural reinforcement; electronic, magnetic, lubricating, and optical devices; chemical sensors and biosensors

Tools for developing autonomous devices that can sense, articulate, communicate, and function as a network, extending human presence beyond the normal senses

Robotics that utilize nanoelectronics, biological sensors, and artificial neural systems

NASA invests up to $1 million per year toward understanding the societal and ethical implications of nanotechnology, with a focus on the area of monitoring human health. University research centers are given opportunities to arrange research by student and postdoctoral fellows, including opportunities to work at NASA centers. One basic NASA nanoscience program in 2003 focused on biomolecular systems research — a joint NASA–National Cancer Institute (NCI) initiative. A second focus is on biotechnology and structural biology. NASA's intent, as noted earlier, is to advance and exploit the zone of convergence of nanotechnology, biotechnology, and information technology.

Collaboration is particularly important for NASA. It recognizes the importance of importing technologies from other federal agencies. Because nanotechnology is in its infancy, the broad spectrum of basic research knowledge performed by other federal agencies would benefit NASA. NASA will concentrate primarily on its unique needs, for example, low-power devices and high-strength materials that can perform with exceptional autonomy in a hostile space environment. A joint program with NCI concerned with noninvasive human health monitoring via identification and detection of molecular signatures resulted from a common interest in this area.

NASA looks to NSF-sponsored work for wide-ranging data arising from fundamental research and emphasizes work in direct support of the Grand Challenge areas
the agency selects for focus in collaboration with DoD (aerospace structural materials,
radiation-tolerant devices, high-resolution imagery), NIH (noninvasive human health
monitoring via identification and detection of molecular signatures, biosensors) and
the U.S. Department of Energy ("lab on a chip"; environmental monitoring).

NASA has significantly increased university participation in nanotechnology
programs by competitively awarding three university research, engineering, and
technology institutes (URETIs) in FY 2003. One area of focus is bionanotechnology
fusion. Each award is about $3 million annually for 5 years, with an option to extend
the award up to an additional 5 years. NASA's Office of Aerospace Technology in
Washington, D.C. established seven URETIs, each in an area of long-term strategic
interest to the agency. The University of California at Los Angeles specializes in
the fusion of bionanotechnology and information technology. Princeton and Texas
A&M Universities specialize in bionanotechnology materials and structures for
aerospace vehicles. The new partnerships give NASA much-needed research assistance in nanotechnology, although its connections with the university research community have declined over the years. All the individual projects within the institutes
have industry as well as university support.

The primary role of each university-based institute is to perform research and
development that both increases fundamental understanding of phenomena and
moves fundamental advances from scientific discovery to basic technology. The
institutes also provide support for undergraduate and graduate students, curriculum
development, personnel exchanges, learning opportunities, and training in advanced
scientific and engineering concepts for the aerospace workforce.

4. *National Institutes of Health*

The National Institutes of Health (NIH) support a diverse range of biomedical
nanotechnology research areas such as:

Disease detection before substantial deterioration of health
Smart MRI contrast agents
Sensors for rapid identification of metabolic disorders and infections
Sensors for susceptibility testing
Implantable devices for real-time monitoring
Implants to replace worn or damaged body parts
Novel bioactive coatings to control interactions with the body
Parts that can integrate with the body for a lifetime
Therapeutic delivery
Addressing issues related to solubility, toxicity, and site-specific delivery
Integrated sensing and dispensing
Gene therapy delivery

The National Institute of Biomedical Imaging and Bioengineering (NIBIB) was
in its formative stages at NIH and became operational in FY 2002. The NIH
Bioengineering Consortium (BECON) coordinates research programs including

nanotechnology research through NIBIB. NIH undertook several nanotechnology-related R&D programs that fell under its FY 2002 research initiative umbrella.

The Genetic Medicine Initiative involves large-scale sequencing to assist in interpreting the human genetic sequence and identifying and characterizing the genes responsible for variations in diseases. An increased investment in nanotechnology research is planned to develop novel revolutionary instruments that can collect DNA sequence variation and gene expression data from individual patients, initially to identify genes involved in causing diseases and later to diagnose the exact form of disease a patient has and guide therapy to treat that patient's disease.

The intent of the Initiative in Clinical Research is to bridge basic discoveries to tomorrow's new treatments, including nanotechnology advances for the development of sensors for disease signatures and diagnoses.

5. Environmental Protection Agency

The Environmental Protection Agency (EPA) recognizes that nanotechnology research has the potential to exert major impacts on the environment via the monitoring and remediation of environmental problems, reductions in emissions from a wide range of sources, and development of new, green processing technologies that minimize the generation of undesirable by-products. Research involving the integration of biological building blocks into synthetic materials and devices will permit the development of more sensitive and smaller sensors.

The goals include improved characterization of environmental problems, significantly reduced environmental impacts from "cleaner" manufacturing approaches, and reduced material and energy use. The potential impacts of nanoparticles related to different applications to human health and the environment have been evaluated. Major nanotechnology-related areas of interest are aerosols, colloids, clean air and water, and measurement and remediation of nanoparticles in air, water, and soil.

The Office of Research and Development (ORD) manages EPA's nanotechnology research. The National Center for Environmental Research (NCER) manages external grant solicitation. In addition, NCER supports a limited number of nanotechnology-based projects through its Small Business Innovation Research (SBIR) program that helps businesses with fewer than 500 employees to develop and commercialize new environmental technologies. The SBIR program links new, cutting-edge, high-risk innovations with EPA programs in water and air pollution control, solid and hazardous waste management, pollution prevention, and environmental monitoring. In-house research facilities include the National Exposure Research Laboratory and the National Risk Management Research Laboratory, and may expand to other ORD laboratories in the future.

In 2003, EPA's research was organized around the risk assessment–risk management paradigm. Research on human health and environmental effects, exposure, and risk assessment gathered to inform decisions on risk management. Research on environmental applications and implications of nanotechnology can be addressed within this framework. Nanotechnology may offer the promise of improved characterization of environmental problems, significantly reduced environmental impacts from "cleaner" manufacturing approaches, and reduced material and energy use.

However, the potential impacts of nanoparticles from different applications on human health and the environment are also being evaluated. Research started in 2002 covers sensors and environmental implications of nanotechnology.

The STAR grant solicitation and SBIR programs are managed by the NCER. In-house research currently includes the National Exposure Research Laboratory and the National Risk Management Research Laboratory, and may expand to other ORD laboratories in the future. EPA has plans to explore collaborations in nanotechnology research with other agencies. In particular, EPA and the Department of Agriculture (USDA) share certain common interests in nanotechnology research, for example, in the areas of biotechnology applications, pesticide monitoring, and food safety.

III. BIOMEDICAL NANOTECHNOLOGY IN EUROPE

A. Introduction

Economically, a sensible strategy for nanotechnology is to focus on niche markets that have no commercially available, cheap, established technological solutions, but which niche markets are relevant for nanotechnology? In Europe, the health care and life science markets may be the best foci for concentration. An early example of a niche market device is the lab-on-a-chip diagnostic technology that is economical and easy to use. The Institute of Nanotechnology in the U.K. is a promoter of this strategy. The German Engineering Society/Technology Center and government studies that prepared the ground for the federal government's competence centers on nanotechnology investigated the potential of nanotechnology in detail for application to various sectors, including medicine, pharmacy, and biology. The competence centers that were set up in 1998 are currently bringing together research organizations, major industries, and SMEs in an effort to stimulate transfers of nanotechnology. This policy follows the example of the bioregions that gave the German biotechnology sector a boost. Other governments and organizations may have their own ideas about potential niche markets to pursue, but it is necessary to bear in mind that technological and economic developments move rapidly and many competitors are working toward the same applications for niche markets and more mature competitive markets.

For the EU and national policy makers, the societal relevance of research is not restricted to economic gains arising from employment and the competitiveness of industries. These decision makers fund research with taxpayers' money and their priorities include better health care, sustainable development, and other benefits. At this stage, one can foresee that nanotechnology is likely to contribute to better medicines and biomedical technologies. It is, however, impossible to quantify the effect.

This section covers biomedical nanotechnology only in the EU research program and in France, Germany, and the U.K. Major nanotechnology initiatives including those aimed at biomedical applications are also ongoing in many other European countries; Switzerland has been the most active.

B. Biomedical Nanotechnology in the EU Research Program

The Sixth Framework Program for Research in the EU spans the period from 2002 through 2006 and highlights nanotechnology as a priority area for European development (see Table 1.2). While the widespread potential applications for nanotechnology indicate that its impact will be felt across virtually the whole program, Priority 3 (nanotechnologies and nanosciences, knowledge-based multifunctional materials, and new production processes and devices) is the main vehicle for research in this area. By bringing together nanotechnologies, materials science, manufacturing, and other technologies based, for example, on biosciences or environmental sciences, work in this area is expected to lead to real breakthroughs and radical innovations in production and consumption patterns. The intention is to promote the transformation of today's traditional industries into a new breed of interdependent high-tech sectors by supporting industry and promoting sustainable development across activities ranging from basic research to product development and across all technical areas from materials science to biotechnology.

The main areas of work identified as suitable and appropriate for funding under Framework 6 include:

1. Mastering processes and developing research tools including self-assembly and biomolecular mechanisms and engines
2. Devising interfaces between biological and nonbiological systems and surface-to-interface engineering for smart coatings
3. Providing engineering support for materials development; designing new materials, for example, biomimetic and self-repairing materials with sustainability
4. Integrating nanotechnologies to improve security and quality of life, especially in the areas of health care and environmental monitoring

Table 1.2 Sixth Framework Funding of European Union

	Million £
Focusing and Integrating Community Research	**13,345**
TP1: Life sciences, genomics, and biotechnology for health	2,255
TP2: Information technologies	3,625
TP3: Nanotechnologies and nanosciences, knowledge-based multifunctional materials and new production processes and devices	1,300
TP4: Aeronautics and space	1,075
TP5: Food quality and safety	685
TP6: Sustainable development, global changes, and ecosystems	2,120
TP7: Citizens and governance in a knowledge-based society	225
Specific activities covering a wider field of research	1,300
Nonnuclear activities of the Joint Research Centre	760
Structuring European Research Area	**2,605**
Strengthening the Foundations of European Research Area	**320**
EURATOM Program	1,230
	17,500

The challenge in the field of materials research is creating smart materials that integrate intelligence, functionality, and autonomy. Smart materials will not only provide innovative answers to existing needs, but will also accelerate the transition from traditional industry to high-tech products and processes. Knowledge-based multifunctional materials were seen as contributors to value-added industries and sustainable development. The strong research in this area should be translated into a competitive advantage for European industries. Another aim of the work package is to promote the uptake of nanotechnology into existing industries including health and medical systems. The priorities include:

New and more sensitive sensors for detection of health and environmental risks
Development of genomics and biotechnology for health
Technology development for exploitation of genetic information, specifically in the area of high precision and sensitivity of functional cell arrays
Improved drug delivery systems

C. France

In France, miniaturization (microsystems) technologies and nanoelectronics are the main foci of nanotechnology research. France has strong nanotechnology research capabilities in the Centre National de la Recherche Scientifique (CNRS) and its universities and a good record in transferring technology from research into the commercial arena. The CNRS and industry jointly fund nano-related research in dozens of laboratories throughout the country. Associated work is conducted by major corporations such as Aventis and Air Liquide. Club Nanotechnologie is a French association that promotes collaborations and exchanges of information.

The jewel in France's research crown is Minatec, the Center for Innovation in Micro- and Nanotechnology, based at the Commissariat à l Énergie Atomique (CEA) Leti facility in Grenoble. The £170 million center aids start-up companies, assists pilot programs for medium-sized companies, and contributes to the R&D programs of large firms. It also brings together CEA Leti and the new Maison des Micro et Nanotechnologies (MMNT) organization. The Grenoble installation will contain resources to promote technical and economic awareness, support start-up operations, and provide offices for national and European networks specializing in micro- and nanotechnology.

1. Government Policies and Initiatives

Since 1999, the French government has been trying to centralize the selection of micro- and nanotechnology and nanostructured materials R&D projects. In recent years micro- and nanotechnology research centers of competence have been coordinated. The Research and Technological Innovation Networks (RRIT) was created by the Ministry of Research and Technology. The RMNT was created in 1999 and provided funding of 10 million annually. Its programs include RNTS (technologies for health) and GenHomme (genomics).

Before 2002 France was a relatively small player in Europe in terms of funding for nanotechnology, but it has substantially increased its investment since 2003 through a coordinated national program considered essential in order to:

1. Develop and upgrade the equipment of the technological centers and clean rooms and open these centers to laboratories and firms
2. Promote the most innovative scientific projects and network the best research centers in the field in order to take advantage of multidisciplinary approaches
3. Encourage mobility among the centers and receive foreign researchers, doctoral candidates, post-doctoral associates, etc.
4. Create new start-ups and SMEs
5. Develop teaching activities at various levels

The national nanosciences program (see Table 1.2) began in 2003 with funding of £15.3 million from MRNT and CNRS and participation from CEA-DSM). Additionally, the Concerted Action for Nanosciences group allocated funding of £12 million for (1) calls for proposals including those in the field of nanobiosciences and (2) integrated projects including architectures of hybrid systems with organic and inorganic nanocomponents. In total, French funding for nanotechnology is approximately £100 million over 3 years, starting in 2003, mainly for five centers:

IEMN, Lille (www.iemn.univ-lille1.fr)
Laboratory for Analysis and Architecture of Systems (LAAS), Toulouse (www.laas.fr)
MINATEC, Grenoble (www.minatec.com)
MINERVE, Paris Sud (www.u-psud.fr/evenement.nsf/projetminerve.html?OpenPage)
LPN,

2. Networks

Twelve nanotechnology networks exist in France according to a survey by the European Commission, including two relevant to biomedical nanotechnology. Biochip Platform Toulouse brings together eight partners in interdisciplinary work to develop new-generation miniaturized biochips in batch production processes. The coordination is handled by the Laboratory for Analysis and Architecture of Systems (LAAS) of the CNRS.

Club Nanotechnologie (www.clubnano.asso.fr) is where researchers and industrialists come together to exchange information on nanotechnology. The chairman is C. Puech, the technical director of Angenieux. Work is undertaken in the areas of metrology, manufacturing, materials, systems, and biotechnology.

D. Germany

Germany's research model for nanotechnology is internationally renowned. Since the end of the 1980s, the German government has supported individual research and development projects in nanotechnology. The German Association of Engineers — the organization responsible for the management of the current national nanotechnology program on behalf of the Ministry for Education and Science,

Research and Technology (BMBF) — produced a strategy document in 1998 titled "Opportunities in the Nanoworld" identifying nanotechnologies critical to the future of industry in Germany. Germany already had a research infrastructure in place, and only modest tweaking was required to meet the new challenges of nanotechnology.

As a result of the strategy document, funding was made available for six competence networks distributed throughout Germany. Additionally, the federal government funds a number of projects in areas such as laser-assisted high-throughput screening of organic and inorganic substances; nanotechnology applications in electronics, medicine, and pharmacy; and nanobiotechnology. The German government provides strong support for nanotechnology. Federal funding for priority nanotechnology research has risen steadily since 1998. Project allocations increased from £27.6 million in 1998 to £88.5 million in 2002 (see Table 1.3).

The nanotechnology research budget for 2003 is £112.1 million, of which £110.6 million is allocated to collaborative research projects involving universities, nonuniversity research institutes, and industries. The remaining £1.5 million is earmarked to fund coordination and improved collaboration within the six virtual nanotechnology networks launched in 1998. Companies participating in collaborative research projects are expected to provide matching funding. In 2001, for example, industry contributed £42 million to R&D collaborations. In terms of technology areas, £9.6 million is available for bionanotechnology research and applications. Funding in Germany is distributed through the country's network of research institutes (Fraunhofer, Max Planck, and Leibniz) and universities. The institutes serve as effective interfaces between basic research and industry, helping to transform basic research into applications. Funding bodies include the federal Ministry of Science (BMBF), research foundation (DFG), the three institutes, the Volkswagen Foundation, and the German states.

Table 1.3 **Annual German Government Spending on Nanotechnology Priority Programs**

Program	Duration	Total Funding (Million £)
Lateral nanostructures	1998–2004	14.32
Nano-optoelectronics	1999–2003	1.53
X-ray technology	1999–2004	5.11
Ultra-thin films	1999–2003	3.07
Functional supramolecular systems	1998–2005	15.34
Nanoanalytics	1997–2005	17.13
Ultraprecision engineering	1999–2004	3.58
Nanobiotechnology	2001–2004[a]	4.09
Nanotechnology competence centers	1998–2003	7.67

[a] Funding for nanobiotechnology projects will be extended beyond 2004; additional funding to be made available.

Source: Faktenbericht Forschung 2002, Federal Ministry of Education and Research, January 2002.

1. Strategy

At a congress held in Bonn on May 6 and 7, 2002, the German Research Minister Edelgard Bulmahn presented the government's strategy on nanotechnology together with an overview of Germany's strengths and research activities in that area. The strategy paper set out measures to promote nanotechnology that encompassed R&D funding schemes, the promotion of young scientists, and public dialogues on opportunities and risks. The overview on Germany's international competitiveness in the area of nanotechnology addressed level of funding, research priorities, and the economic potential of nanotechnology in Germany. Total expenditures on nanotechnology research and development in Germany in 2001 totalled £217.3 million. This amount includes £153.1 million from the public sector — both institutional and project funding — and £64.2 million from industry sources.

The federal government recognizes the importance of nanotechnologies as key enabling technologies for a wide range of sectors including biotechnology and analytics. It has therefore made nanotechnology a key research priority and supports the exploitation of its commercial and job-creating potential and wider dialogues on the opportunities and risks. BMBF published a strategy titled "Nanotechnology in Deutschland: Strategische Neuausrichtung" It also produced an overview of Germany's R&D priorities and strengths in different fields of nanotechnology — "Nanotechnologie in Deutschland: Standortbestimmung." Both documents have been published in German and are available on the Internet at www.bmbf.de. Information about the virtual nanotechnology clusters in Germany is available at www.nanonet.de (including English language information) or via the links listed above. The web pages list individual members in each cluster. BMBF continually sets priorities in research programs within the framework of nanotechnology (since 1999) and nanobiotechnology (since 2000):

Materials research (nanomaterials, analytics, layers)
Microsystems technology (sensoric layers)
Biotechnology (drug delivery systems, data processing with biomolecules)

2. Nanobiotechnology

In 2000, BMBF launched its Nanobiotechnology (NB) Program dedicated to the funding of multidisciplinary research projects related to:

Development of analytical and characterization processes with resolution in the nanometer range
Establishment of manipulation techniques for biological and functionally analogous biochemical objects
Development of reaction techniques for the analysis of structure–activity relationships
Use of biological self-assembly mechanisms for the development of functional layers and surfaces
Design and application of cellular and molecular tools and machines

The funding activity is a joint initiative between BMBF's Physical and Chemical Technology Program and its Biotechnology Framework Program. A total of £50 million has been earmarked for 6 years. It complements current funding activities in the areas of nanotechnology, proteomics, material sciences, and others. The major goals of the NB program are:

Rapid transfer of biological expertise into nanotechnology
Use of biological nano-sized objects in technical systems
Effective exploitation of nanotechnology in biotechnology and medicine

Because applications from NB are varied, the projects involved relate to a wide range of research areas, for example, (1) application of nanoparticles in drug delivery and diagnostic systems, (2) use of nanostructured biological surfaces in technical systems, for example, data storage, and (3) development of biosensors and micro-arrays. Further information is available at www.bmbf.de and www.nanobio.de.

3. Competence Networks

Additional biomedical nanotechnology research is funded through several other competence networks. One network is Nanotechnology: Functionality through Chemistry. In most industrialized countries, the application of chemical principles to prepare nanostructured materials is increasing in fields such as pharmaceuticals, dispersion paints, optimization of catalysts and glues, and lack and smear processes. Eighteen universities, 23 research centers, 50 small and medium enterprises, 15 large companies, and 7 risk capital groups have joined in a virtual center of competence that covers the whole value chain (education, research, development, production, and marketing).

Nanobionet is another competence network. Its aim is to develop applications of nanobiotechnology in the fields of pharmacy, new medicine, artificial photosynthesis, antibacterial coatings, and functional textiles. Universities and 50 companies in the Saarland, Rheinhessen, and Pfalz regions in Southwest Germany are collaborating. The Münster Bioanalysis Society is a network of business, science, and government entities that focuses on nanobioanalytic activities in the Münster region. The national competence networks are intended to enable domestic manufacturers to commercial-ize nanotechnology. Large companies collaborate actively in the networks and are very aware of new developments. Another aim is to create jobs in innovative sectors in Germany and protect the existing ones in a globally competitive market. Germany sees important opportunities and has strengths in nanotechnology applications for electronics and data storage systems, chemicals and materials, optics, vehicle tech-nology and mechanical engineering, and microscopy and analytics.

In other important nanotechnology applications, for example, nanobiotechnology and display technology, Germany is perceived as lagging behind its main competi-tors. About two thirds of research funding is strategically directed, while the final third is opportunistic. The emphasis is on applied research without neglecting more speculative research.

4. Research Centers

Germany has a very large nonuniversity research infrastructure. In addition to research activities at universities and institutes attached to universities, research is undertaken in institutes of the Max Planck Society (79 institutes), the Fraunhofer Society (48 institutes), the Leibniz Association (78 institutes), and the Helmholtz Association (16 national science centers). The federal and state or municipal governments fund these research organizations jointly with the intent to clearly delineate the functions of these organizations. The Max Planck Society is devoted to pure research. The Fraunhofer group pursues applications-oriented research, and the university spin-out institutes mainly focus on specific commercial areas. This distinction is blurring slightly because of industry demands for access to expertise from the Max Planck institutes.

The presence of a strong and comprehensive research infrastructure has made it simpler to supply additional funding to support specific needs in emerging areas such as nanotechnology. The government is sending an increasingly powerful message that the research is required to yield products and jobs. This represents a fundamental shift in the attitude of German researchers toward commercialization, although failure in business remains unacceptable.

a. CAESAR

The Center for Advanced European Studies and Research (CAESAR) is a scientific research center funded as part of a compensation package for the move of the federal government from Bonn to Berlin. The operational structure described below is interesting and novel; research is firmly targeted at short-term commercial applications. Nanotechnology is considered a major research focus at CAESAR under:

Dr. Jorgen Refresh (structure, mission, transfer policy)
PD Dr. Michael Mosque (thin adaptive films)
PD Dr. Elkhart Quanta (smart materials)
Dr. Daniel Hoffmann (protein folding)

CAESAR was inaugurated in 1995 as a new type of research center with the aim of catalyzing scientific and economic activities and creating jobs. It is a private, nonprofit research institute that carries out research at the interface of information technology, physics, materials science, chemistry, biology, and medicine. The goal of each research project is to create marketable innovations that lead to the establishment of start-up companies or industrial exploitation.

This goal is reached by (1) pursuing multidisciplinary time-limited research projects, (2) assembling temporary teams of researchers employed by CAESAR and by other research organizations and industry, (3) developing new mechanisms for commercialization, including the substantial support of start-up companies, and (4) serving as a nucleus for cooperative activities and a focal point for local knowledge networks.

The operational structure is project-oriented, with small groups of about five scientists undertaking fixed period (say, 5 years) tasks. At the end of the period, they leave to work elsewhere. The CAESAR organization works cooperatively with local institutes and universities.

The research is focused on (1) nanotechnology and materials science, (2) biological and electronic systems, and (3) ergonomics in communications and surgery. Since its inception, CAESAR has launched 4 start-up companies and 20 industrial collaborations aimed at new product development. In nanotechnology, automotive applications have been identified for thin film sensors.

b. Charité

Charité is Europe's largest university clinic and medical faculty based at three sites: Virchow-Klinikum, Charité Mitte, and Berlin Buch. The biomedical nanotechnology group evolved from the radiology department in Virchow. Led by Dr. Jordan, the group recently developed a method of introducing colloidal dispersions of super paramagnetic biocompatible iron oxide nanoparticles into tumors. This work led to the formation of two spin-off companies, MFH GmbH and MagForce Applications GmbH.

c. Institute for New Materials

The Institute for New Materials (INM) is a model for a research and development institute that achieved a world class reputation for innovation in new materials in a relatively short time. Many of its innovations involved nanoscale technologies. The INM, unique in the world of German materials research, was founded with the long-term R&D objective of introducing new high-tech materials on a commercial scale. Highly innovative high-risk long-term basic research has been funded with the aim of reducing the 10 to 15 years required to develop new material technologies from idea to marketplace. Products and processes nearing commercial application are developed in cooperation with industrial partners that also provide the necessary financing. This successful approach has enabled the INM to expand quickly into a research institute with 250 employees housed in a new 10,000-square-meter facility and a turnover greater than £15 million.

To achieve the greatest possible variety of high-tech materials, the INM adopted the strategy of integrating inorganic synthesis chemistry with chemical nanotechnology. This combination has been the key to a whole new world of materials. The INM was one of the first research institutes to consistently use chemical synthesis including the sol–gel process as the basis for manufacturing materials with the assistance of nanotechnology.

The INM enjoys considerable national and international commercial collaboration and is a key player in several networks. It is a member of the Centre of Excellence in Nanotechnology, a network involving 65 industries and 42 institutes. The INM also runs conferences and workshops on a variety of materials-related topics. It is one of the centers of competence created by the government; it has a spin-off company called Nanogate; and it runs a joint venture with TNO, the Dutch technology organization.

In the nanotechnology area, the INM is developing sol–gel technology into ormocils and ormocers using interpenetrating networks of inorganic and organic molecular structures to provide functional coatings. Chemical nanotechnology (a combination of organic and inorganic colloidal chemistry) is used to combine a sol–gel or polymer matrix with nanomers — external ceramic, metallic, or semi-conductor particles — to achieve a range of properties. These nanomers can be single-component or multicomponent (alloys, core shells) structures. The particles can be closely packaged in substrates, widely dispersed, or function as nanopowders. Resultant materials can be transparent composites with advanced properties (hardness, scratch resistance, durability, and others). Viscosity can be controlled. The technique is leading to the development of new binding agents, transparent fillers with specific shrinkage, thermal expansion, and thermal conductivity features.

d. Institute of Microtechnology Mainz

The Institute of Microtechnology Mainz (IMM) in Germany has 160 staff members. It specializes in microfabrication methods including LIGA techniques, ultraviolet lithography, thin-film technology, ultraprecision engineering, laser micromachining, and micro-EDM that have applications in fields such as microreactors, biomedical devices, microoptics, sensors, and actuators. Its nanotechnology research concentrates on the development of tools for scanning probe microscopy.

e. Max Planck Institute of Colloids and Interfaces

The Max Planck Society for the Advancement of the Sciences is an umbrella of 81 independent institutes that focus on new fundamental research that cannot be accommodated easily within a university environment due to its multidisciplinary nature or requirements for staff and/or facilities. The Max Planck Institute of Colloids and Interfaces is an outcome of reunification. It was founded in 1993 as one of the first Max Planck Institutes of East Germany. It brought together the three former German Democratic Republic institutes of polymer, organic, and physical chemistry. The aim of the new institute was to build a multidisciplinary research base that looked to the future, attracting talent from different backgrounds and integrating existing staff from both East and West Germany.

Although the institute's stated objective and desire is fundamental research, it finds it increasingly difficult to maintain this limitation. Some industrial cooperation exists, for example with L'Oreal, BASF, and Roche which together provide a surprising 40% of the institute's funding. Industry continues to exert pressure on the institute to form more partnerships; this evidences growing industrial interest in the topics studied. The institute is now at the stage where it must field requests from industry in order to concentrate on its own pure research agenda. However, the commercial potential of research outcomes is not ignored, and several applications are currently in the process of commercialization. Researchers and their activities include:

Dr. Helmut Culfen: Biomimetic mineralization, fractionating colloid analytics, filament growth forming neuron-like networks

Dr. Katharina Landfester: Mini-emulsion polymerization, particle synthesis within micelles, nanocapsules

Dr. Roland Netz: Theoretical approaches to nanoscopic systems

Other research areas are nanoparticle chemistry, scale-up of nanoparticle production, quantum dots, phosphors, biolabeling, bioimaging, cell death, directed deposition, security products, inks, and heterogeneous and homogeneous catalysts. Future projects of the institute will focus on artificial cells with specific reference to membrane and interface functions, theories of biomimetic systems, new concepts in colloid chemistry, compartmentalization of biomimetic chaperone systems, and nanocrystallinity. Staff scientists lead small, largely independent groups. Good interdisciplinary contacts exist among the various project groups at the institute, and strong external links exist through joint projects with the four Berlin universities, the neutron reaction source at the Hahn–Meitner Institute, and the synchrotron radiation facility known as BESSY.

The institutes derive particular benefits for developing leading-edge research based on the way the funding system operates for the Max Planck institutes. The government provides funding and allows each institute to set its own research agenda. The institutes are under no great pressure to find commercial partners. The current trend in Germany is toward funding larger projects with budgets of £5 million to £25 million. A serious problem is finding enough physics and chemistry students; many are now recruited from Eastern Europe and China.

E. United Kingdom

1. Introduction

The United Kingdom showed an early interest in nanotechnology. Its DTI National Initiative on Nanotechnology (NION) was announced in 1986, followed in 1998 by the 4-year LINK Nanotechnology program. The final funding for LINK projects was handed over in 1996. After that, the United Kingdom had no national strategy for nanotechnology, although dispersed research involving nanoscale science continued to be funded. In 1997, the Institute of Nanotechnology, a registered charity, was created to fill the gap and act as a focus of interest in nanotechnology throughout the United Kingdom. The institute grew out of the Centre for Nanotechnology which received a small amount of funding under NION to raise awareness of nanotechnology and its applications.

Oxford and Cambridge lead the way in England in terms of nanotechnology research and spinning out companies, but the country has a number of other significant centers and universities, with over 1,100 researchers nationwide. Imperial College London recently established the £9 million London Centre for Nanotechnology, and major centers have been established in Birmingham and Newcastle. Many universities have set up the interdisciplinary infrastructures required for nanotechnology research. Master's programs now exist at Leeds, Sheffield, and Cranfield. The University of Sussex started offering nanotechnology degrees in 2003.

Since 2000, government support for nanotechnology research in universities has increased significantly. The new innovation centers for studying microsystems and nanotechnology have been set up at the Universities of Newcastle and Durham. Two interdisciplinary research collaborations (IRCs) split £18 million in funding. The first, focusing on the biological aspects of nanotechnology, is led by Oxford University. Nanotechnology research in the United Kingdom is becoming more commercial in its outlook, and the government's nanotechnology initiative will seek to further support this development.

In Summer of 2001, Lord Sainsbury, the Parliamentary Under-Secretary of State for Science and Innovation, announced that nanotechnology would play an important role in new initiatives. It was a prime candidate to participate in the £41 million basic technology program announced under the government's spending review. This program provides funding for high-risk research that may result in some new disruptive technological development. (A disruptive technology totally removes its predecessor from the scene — for example, compact disks replaced long-playing records). The program is only open to higher education institutions.

In addition to the £41 million for research, the government also introduced a new 3-year, £25 million program aimed at helping businesses commercialize key technologies emerging from the basic technologies program. Nanotechnology is also one of the four key research priorities in the third round of the Foresight Link Awards. The awards have a £15 million budget.

2. Interdisciplinary Research Collaborations

In 2001, £18 million was awarded for two IRCs in nanotechnology to consortia headed by Oxford and Cambridge Universities after their proposals were chosen from a total of 16. Funds for these collaborations have become available through three government research councils (EPSRC, BBSRC, and MRC) along with the Ministry of Defense. The awards represent the government's largest commitment to nanotechnology to date. After 6 years, the IRCs will revert to conventional means of support.

The essential elements of an IRC are (1) a critical mass of researchers, (2) a concentration of advanced instrumentation, and (3) excellent multidisciplinary research and training opportunities. IRCs are expected to nurture the "revolutionary" aspects of nanotechnology and provide a firm foundation for "evolutionary" studies building on established technologies. Industry has a critical role in further defining the scope of the IRC.

The nanobiotechnology IRC is headed by Oxford University with participation of the Universities of Glasgow and York and the National Institute for Medical Research. This collaboration also involves links with the Universities of Cambridge, Nottingham, and Southampton. The consortium is directed by Professor John Ryan who heads Condensed Matter Physics and the Physics Department. The Glasgow group, led by Professor Jon Cooper and a team of six other academics, seeks to combine expertise in nanotechnology, lab-on-a-chip, and biosensor devices in order to develop a series of extremely sensitive tools that will enable biologists to manipulate and measure single biological molecules (see below). This will help determine how the genetic code controls the behavior of cells and how the activities of drugs control cell metabolism.

Molecular machines — These machines are proteins that convert electrochemical energy generated across a membrane into external mechanical work. They are responsible for a wide variety of functions from muscle contraction to cell locomotion, copying and processing DNA, movement of chromosomes, cellular division, movement of neurotransmitter-containing vesicles, and production of adenosine triphosphate (ATP). The mechanical properties of molecular motors can be considered in terms of rectifying thermal ratchets and impedance-matching lever systems that couple enzyme-active sites to external loads. For many systems, it is now possible to reconstitute their functions using purified proteins and to observe and measure the forces and movements that they produce during a single chemical cycle. In other words, we can measure the mechanochemical processes that take place at the level of a single molecule. Furthermore, "man-made" molecular motors now in development are based either on hybrid constructions of existing rotary and linear biological motors or produced from man-made materials and based on molecular motor design principles.

Functional membrane proteins — The fact that 15 to 30% of all genes code for membrane proteins provides evidence of their immense biological importance. Membrane proteins include ion channels (that enable rapid yet selective flux of ions across membranes), hormone receptors (that may be viewed as molecular triggers and amplifiers), and photoreceptors (protein molecules switched between two conformational states by the absorption of a single photon of visible light). The structures of these proteins were poorly described structurally until recent advances in structural biology (x-ray diffraction and solid state nuclear magnetic resonance [NMR]) greatly improved our understanding of membrane protein structure. It is now possible to explore their structure–function relationships at atomic resolution level and exploit their unique dynamic properties.

Bionanoelectronics and photonics — One key issue of all aspects of bionanoelectronics is the attachment of biomolecules to surfaces. This is a pervasive problem in designing most sensors and investigating cell–substrate interactions, biocompatibility, and the realization of DNA and other biopolymer sequencing devices. Nanofabrication methods will be used to produce surfaces patterned both topographically and molecularly at the nanoscale level. Macromolecules can be assembled into two- and three-dimensional constructs.

Electronic circuits and networks — The construction of electronic circuits and networks is one of the grand challenges of bionanotechnology. Carbon nanotubes and DNA oligomers such as double-stranded poly(G)–poly(C) are possible candidate molecular wires. Nanotube electronic circuits may be constructed using atomic force microscopy (AFM) manipulation; charge transfer in DNA oligomers can be studied using nanostructured electrical contact arrays and ultrafast optical techniques. DNA has important additional advantages in that networks may be produced by self-assembly.

Photonic applications — The classic bacteriorhodopsin (bR) membrane protein has been shown to be an effective material for photonic applications such as optically addressable spatial light modulators, holographic memories, and sensors. The photosynthetic reaction center is only 5 nm in size and behaves as a nanometer diode. Its integration with nanotubes and nanometer electrodes will provide unique opportunities for bioelectronic logic devices, transducers, photovoltaic cells, memories, and sensors.

Single-molecule experimental techniques to be employed extensively in the IRC program include AFM, scanning tunneling microscopy (STM), optical and dielectric traps ("tweezers"), scanning near-field optical microscopy (SNOM), fluorescence resonant energy transfer (FRET), and single-channel patch clamping.

The second IRC will concentrate on the physics of nanotechnology and is led by Cambridge University, with participation by University College London and the University of Bristol. The consortium is directed by Prof. Mark Welland, head of the Nanoscale Science Laboratory in the Department of Engineering at Cambridge. The other six investigators are Prof. Richard Friend (Cambridge, Physics), Dr. Mark Blamire (Cambridge, Materials Science and Metallurgy), Prof. Chris Dobson (Cambridge, Chemistry), Prof. Mervyn Miles (Bristol, Physics), Dr. Andrew Fisher (University College London, Physics), and Prof. Michael Horton (University College London, Medicine).

The IRC's activities will focus on the general themes of fabrication and organization of molecular structures. Material systems the study intends to cover include molecular materials for electronics and photonics, self-assembly approaches to well-defined structures including the investigation of fibril structures in proteins and polypeptides, controlled cell growth from substrates for tissue engineering, and the creation of natural biosensors.

Newcastle University was awarded £4.6 million in 2001 to create a university innovation center (UIC) for nanotechnology. This funding partly supports a high-technology cluster development initiative to build on nanoscale science and technology activities at the five universities in northeast England, and includes support from the private sector and the One NorthEast regional development agency. The regional portfolio encompasses surface engineering (Northumbria), chemical and biological sensors (Sunderland and Teeside), molecular electronics (Durham), and biomedical nanotechnology (Newcastle). Together the UIC and the International Centre for Life in Newcastle that services the biotechnology sector will act as a cross-sector driver for regional high technology-based cluster development.

On July 2, 2003, Lord Sainsbury announced funding of £90 million over the next 6 years to help United Kingdom industry harness the commercial opportunities offered by nanotechnology.

IV. JAPAN

A. Introduction

Government agencies and large corporations are the main sources of funding for nanotechnology in Japan. Small- and medium-sized companies play only minor roles. Research activities are generally handled by relatively large industrial, government, and academic laboratories. According to a report by the *Journal of Japanese Trade and Industry*, the Japanese government views the successful development of nanotechnology as the key to the restoration of the Japanese economy.

Most of Japan's nanotechnology funding, supported by a number of agencies since the 1980s, was oriented toward studying nanoscale phenomena in semiconductor materials or developing new materials. Japan became involved from an early stage in advanced nanotechnology research that led to a nanomechanism project in 1985, the discovery of carbon nanotubes by Dr. Iijima Sumio in 1991, and atom technology in 1992. Policymakers have been strengthening research capacity at the interface of life sciences and nanotechnology. Research centers in the public sector are increasingly offering positions for scientists specializing in nanobiology; funding schemes for young scientists in nanobiotechnology are emerging.

Japan was spurred into action by the NNI in the United States. The Japanese government founded the Expert Group on Nanotechnology under the Japan Federation of Economic Organizations' (Keidanren) Committee on Industrial Technology. Japan targeted nanotechnology as one of four priorities in its fiscal 2001 science and technology budget. The Council for Science and Technology Policy, chaired by Prime Minister Junichiro Koizumi, decided that life science, information and telecommunications, the environment, and nanotechnology together would reinvigorate the Japanese economy.

Despite pressure to rationalize public expenditures, the Japanese government continues to invest heavily in nanotechnology. The fields of materials development and nanoscale fabrication continue to occupy the spotlight. Nanotechnology is now the key priority under Japan's second basic S&T plan and gained a substantial increase in funding from the central government for fiscal year 2002. The total nanotechnology budget will be around £74.6 billion, with most of the amount committed from the Ministry for Education, Science, and Technology (MEXT) and the Ministry of Economy, Trade, and Industry (METI). Table 1.4 summarizes nanotechnology funding in Japan for 2001–2003.

B. Government Policies and Initiatives

In 1995, the Japanese Diet enacted the Science and Technology Basic Law requiring the government to develop and implement two successive 5-year basic science and technology plans. The first plan became effective on April 1, 1996 and was completed on March 31, 2001. The government spent £17 trillion for R&D under the first basic plan. The second plan extends from April 1, 2001 through March 31, 2006 and the government expects to invest £24 trillion in S&T (assuming 1% of the Gross Domestic Product and nominal GDP growth of 3.5% per year). The highest priorities under the second plan are:

Life Sciences: prevention and treatment of diseases; elimination of starvation
Environment and human health: preservation of the environment to maintain our basic
  existence

Table 1.4 Japan Nanotechnology Funding Summary, 2001 through 2003 (units in 100m yen)

Fund Agency	2001	2002	2002 Supplementary Budget	2003 (proposed)
Ministry of Education, Culture, Sports, Science and Technology	21	98	172	106
Ministry of Economy, Trade and Industry	195	312	50	372
Ministry of Public Management, Home Affairs, Posts and Telecommunications	3	19	0	19
Ministry of Health, Labor and Welfare	0	12	0	18
Ministry of Agriculture, Forestry, Fisheries of Japan	1	2	0	2
National Labs Budget	35	137	0	137
Competitive Research Grant	353	388	0	388
TOTAL	606	969	223	1042

Note: Figures represent 100 millions. Nanotechnology funding in Japan includes MEMS and Semiconductor Nanoelectronics programs. The Japanese financial year for 2003 started April 1.

Source: Council for Science and Technology Policy (CSTP), Cabinet Office, Japan.

In 2002, MEXT launched the Nanotechnology Researcher Network Center of Japan (NRNCJ). The center provides core facility services, information services, and technology transfer support for Japanese researchers in nanotechnology and nanoscience.

C. Support and Development

The Japanese government has a broad-based systematic plan to promote and support the development of nanotechnology. Projects are classified into four categories:

Basic research — The focus is on the development of a basic understanding of nanotechnology along with the development of nanoscale particles and nanostructure materials based on extensive exploratory and in-depth research in physics, chemistry, and biology, and development of new theories and methods of modeling, simulation, and analysis.

Generic technologies — Research topics are nanoanalyses, nanofabrication, and nanosimulations.

Challenge-type projects — The goal is to focus on R&D projects that will create fundamental and revolutionary technologies to support industry in the next 10 to 20 years. An example of an R&D project in this category is development of biomaterials and biosystems and technologies for medical and health care use by fostering interdisciplinary projects involving biotechnology and nanosystems.

Flagship-type projects — The R&D will focus on areas of technology that have practical applications and will produce economic impacts within 5 to 10 years. An example of such a project is next-generation semiconductor technology using the conventional top-down approach.

D. Nanotechnology Virtual Laboratory

The research in this nanotechnology virtual laboratory is carried out by small interdisciplinary teams across a number of strategic nanoscience fields. Research targets technologies expected to emerge into the marketplace in the next 10 to 20 years, for example, biodevices, nanocomposites, drug delivery systems (DDSs), and programmed self-assembling molecules.

E. Nanotechnology Project of Ministry of Health, Labor, and Welfare

The Ministry for Health, Labor and Welfare (MHLW) launched its first nano-technology-related project in 2003, with around £1.4 billion committed to nanomed-icine. Specifically, this project will cover analysis of proteins and the development of miniaturized surgical equipment and drug delivery systems. About half the funding will be directed to the National Medical Center; the remainder will be available to other researchers via a public tendering process.

V. CONCLUSION

This chapter described the development of nanotechnology programs in the United States, the European Union, certain large European countries, and Japan since the 1990s. We outlined the major research priorities and described the institutes specializing in biomedical nanotechnology in these countries. The remainder of the book will cover nanotechnology developments related to drug delivery, diagnostics, prostheses and implants, and biodefense issues; it will also examine socioeconomic aspects and health risks.

CHAPTER 2

Nanotechnology and Trends in Drug Delivery Systems with Self-Assembled Carriers

Kenji Yamamoto

CONTENTS

I. INTRODUCTION

This chapter describes the applications of nanotechnology in drug delivery systems with self-assembled drug carriers. The development of this technology since the 1980s is described and the different technologies applied are explained. These types of drug delivery systems are promising for cancer therapy applications. Present chemotherapy systems cause severe side effects. Targeted drug delivery systems can help reduce the side effects because they deliver medication to cancerous cells rather than spread it via the circulatory system. Nanodrug delivery is becoming a very large and fast-moving field. For that reason, this chapter focuses on certain elements and explains them in depth rather than attempting to cover every aspect of the subject briefly.

II. DRUG DELIVERY SYSTEMS SINCE THE 1980s

The technology involved in a drug delivery system can be classified into three fields: releasing technology, targeting technology, and controlled membrane transport. The length of the holding time of an efficient concentration of a drug depends on the half-life of the drug inside the body, as is also true for nuclear molecules. Holding time depends on the velocity of the inactivation of the drug inside the body or the velocity of releasing the drug outside the body. In order to retain efficient concentration inside the body for longer times, we have to prescribe a higher dose.

An ideal drug for avoiding side effects would have the ability to raise its concentration up to the efficient level immediately after the dose is given, hold the level for a constant period to allow the drug to do its work, and return to the original level soon after the treatment period so as not to interfere with the subsequent dose. A suitable releasing technology that achieves these purposes would be desirable. The three controlled-release technologies available at present are the (1) pulse-release — a constant amount of drug is released at a constant time interval; (2) feedback-release — drug is released on command from a physical signal; and (3) constant-release — drug is released at a constant rate. Two types of targeting technologies are available. One is the active type that utilizes a signal peptide, the antigen–antibody reaction, and the receptor-ligand. The other type is passive and utilizes the enhanced permeation and retention (EPR) effect near a malignant tumor organ.[1]

For controlled membrane transport, we can combine specific physical stimulations and pro-drug technology to increase efficiency. The pro-drug technology is described briefly as follows. A drug that is less efficient at the point of membrane transportation is modified chemically so that it can be transported more easily across the membrane. After transportation, the modified drug returns to its initial state or changes into derivatives that produce the intended activity inside the tumor.

One major technology is the enhanced permeation and retention effect discovered by H. Maeda's group in 1986.[1] Inside the cancerous organ, macromolecules easily permeate the newly manufactured blood vessels. At the same time, macromolecules are hardly released from the organ through the lymphatic vessels. As a result, the macromolecules are retained inside the cancerous organ. During the past few years, this finding allowed major progress in targeting technology against solid tumors.

Another example is poly(styrene-co-maleyl-half-n-butylate) neocarzinostatin (SMANCS) technology. SMANCS (molecular weight [MW] 15,000) is a supermolecule consisting of neocarzinostatin (NCS; MW,1,100) covered with a styrene–maleic acid co-polymer discovered in 1978.[2] In 1982, SMANCS covered with iodized poppyseed oil was first injected through a human hepatic artery to induce an embolism that was necessary to retain the drug for a time.[3] The human liver has four blood vessels, two of which transport blood into the liver and two that remove it from the liver. The two incoming vessels are the portal vein and the hepatic artery. The portal vein contains a high concentration of nutrient substances and a low concentration of oxygen. The hepatic artery contains a low concentration of nutrient substances and a high concentration of oxygen. A normal hepatic cell is supported mainly by the blood from the portal vein. A hepatic cancer cell (HCC) is supported by the hepatic artery. An HCC requires active aerobic respiration and cannot survive under a low partial pressure of oxygen. In practice, a cancer cell stops growing through the embolization of the hepatic vessel located upstream of the tumor and dies via the release of a high concentration of an anti-cancer drug from the SMANCS particles retained in the tumor.

Other techniques devised to deliver drugs via nanotechnology include a system by Duncan based on polyethylene glycol (PEG) methacrylate tagged with an anti-cancer drug through a peptide bond.[4] Another drug delivery system is based on macromolecules with dendritic polymers conjugated with cisplatin–methotrexate for the treatment of cancer by Frechet's group.[5] Baker's group produced a drug delivery system based on sialic acid for the prevention of influenza pneumonitis.[6] Another drug delivery system reported by N. Yui is based on a supramolecule pro-drug technique that uses thermally switchable polyrotaxane.[7]

Another application for a drug delivery system is as a carrier of gene therapy. One established method of gene therapy uses a virus to deliver the genes necessary for healing the patient into target cells. Recently, Cavazzana-Calvo's group[8] reported that the inappropriate insertion of such a retroviral vector near the protooncogene *LMO2* promoter led to uncontrolled clonal proliferation of mature T cells in the presence of the retrovirus vector.[8] To avoid such a risk caused by a virus vector, a gene delivery system (GDS) with a nanocarrier would be a possible method of therapy. We found several references to such nano-gene delivery systems, as follows. A nonviral gene transfer system based on a block polymer was developed by K. Kataoka.[9] A. Florence et al. devised self-assembled dendritic polymers conjugated with DNA[10]; and a system involving a membrane fusion liposome Sendai virus protein was proposed by Eguchi et al.[11]

A. Government Funding for Nanodrug Delivery Systems

Until recently, large-scale research and development in nanotechnology were activities pursued by industries and national programs of governments of many countries including the United States, the European Union and its member states, and Japan. National budgets have been invested in research and the development related to drug delivery systems. The National Nanotechnology Initiative (NNI) in the United States, the Sixth Framework Program for Research and Technological Development of the European Union, and the Council for Science and Technology Policy of the Cabinet Office in Japan are examples of national efforts targeted toward drug delivery systems involving nanotechnology.

Self-assembly is one of the common processing nanotechnology methods for producing functional nanometer-sized particles (supermolecules). This review focuses on the development of nanotechnology for applications in drug delivery systems, particularly the self-assembled supermolecules.

III. CHEMICAL SYSTEM ENGINEERING AND NANOTECHNOLOGY

Some of the terms to be used in this chapter should be defined more clearly. A chemical system is defined as a set of chemical elements that have complex relations with each other and as a whole perform certain comprehensive functions. Chemical system engineering is defined as a group of thoughts, theories, and ways to utilize chemical systems to benefit human beings. Our definition of a *chemical system* is not restricted only to chemical materials such as compounds and assembled particles. We would also extend this definition to biological entities including viruses, cells, and bodies, all of which consist of chemical elements. A complete biological entity also performed certain functions as a living organism.

By using the broad definition, the phenomena observed in the systems described below can be represented with the fundamental equations of the systems of particles. These equations can cover areas as diverse as the diffusion reaction function, the systems of links among living bodies, and even analyses of social relationships.

The pattern formations of bacterial colonies such as *Escherichia coli* and *paeni-bacillus dendritiformis* were analyzed with nonlinear differential equations.[12] In *Bacillus subtilis*, the phase transition of the morphology was induced by the con-centrations of the nutrients[13] and analyzed by using the chemical system approach. One of the colony patterns was solved with nonlinear differential equations; the cell was regarded as a self-growing particle assembled from the chemical compounds in the medium.[14,15]

We cannot say that we can analyze the colony patterns of microorganisms by means of the genome project or the post-genome project currently in progress. These programs are concerned with sequential information and not chemical pattern for-mations such as the "Turing patterns" of Belousov-Zhabotinsky reaction. The view-point described above can be considered an important and useful approach not only for chemical system engineering, but also for the understanding of life.

We define a supermolecule as a particle consisting of a set of chemical elements in which any element has some complex relations with other elements. A whole supermolecule can perform some comprehensive functions. For example, a red blood cell carrying oxygen could be thought of as a particle that contains a huge amount of hemoglobin. The outer shell (cell membrane) consists of a lipid bilayer. The functions of a supermolecule are not limited to those of the assembly of individual molecules; a supermolecule can function as a whole.

We define nanotechnology as a system of thoughts, theories, and methods that allow us to design a supermolecule, to realize it in production, and utilize it for industrial manufacturing and in daily life. One object of nanotechnology is the design and production of supermolecules regardless of their size.

Finally, bionanotechnology is very much like nanotechnology except that the supermolecule in bionanotechnology includes not only the function but also the information of the whole particle. For example, consider a filler particle for a liquid crystal display. The filler nanoparticle should be designed to be small enough to move efficiently through the pathway. After the particle reaches its destination and releases information indicating that the place has been reached, the surface arms that are designed to stretch out and stack fix the parts of the liquid crystals tightly. In bionanotechnology as defined above, we are developing a particle that will contain such installed functions as the sensing of status, exchange of information, and making a precise decision related to the functional proceedings in the same way a living organism reacts in nature.

IV. TOWARD DEVELOPMENT OF DRUG DELIVERY SYSTEMS WITH BIONANOTECHNOLOGY

A. Self-Assembly and Self-Organization

Two methods exist for processing material as shown in Figure 2.1. The top-down method is the manufacturing of functional end products from a bulk material. The second method involves the design and manufacture of a fundamental unit after which a functional product is assembled from the set of units; this is known as the bottom-up method.[16] The cell utilizes this type of self-assembly technology to make certain materials in order to stay alive. One example is the bacterial flagellar protofilament.[17] The unit is designed to be assembled by itself to facilitate the process of the production of nanostructures (nanotubes and nanovesicles).[18–20]

The idea of self-organization is similar to that of self-assembly. Through the self-assembly method, a product grows layer by layer with a high degree of equilibrium (Figure 2.2).

Conversely, a product produced through the self-organization method is made with a high degree of nonequilibrium. In this method, the product is made all at once from the start instead of being assembled one layer at a time. An end product made with the desired functional structure by this method does not have a minimum of free energy, but has a minimum loss of entropy. The bottom-up method has another superior characteristic. As the end product is made from the fundamental units by

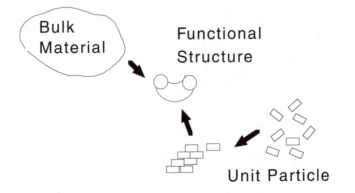

Top Down *vs.* Bottom Up

Bulk Material

Functional Structure

Unit Particle

Figure 2.1 Top-down and bottom-up methods.

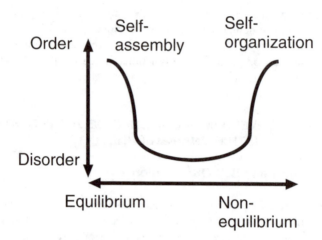

Figure 2.2   Self-assembly and self-organization.

self-assembly, a small change of the fundamental units can lead to a significant change in the character and the function of the assembled final product.[21]

The reverse question will arise: how can we design a fundamental starting material unit for making a final product that has different characteristics from the original one? For example, collagen is a biomaterial made by animals and plants that has a mesh structure and is used in many different ways for biological and medical applications. We can design an oligopeptide for the processing of the product through the self-assembly method, as for the substitution of the collagen. The two oligopeptides discussed below are among the examples for such use.

RADA and EAKA tetramers — The common structures of these two fundamental units consist of positively charged and negatively charged amino acids positioned alternatively among the hydrophobic amino acids. The unit molecules hold beta structures and self-assemble each other by intermolecular beta–beta interactions. The assembled products are known to grow into fiber-like structures[22] that are known to hold a characteristic three-dimensional structure and can be used as a substitution for collagen on a cell culture dish.[23]

One of the incentives that promotes the development of a substitute for collagen is that the collagen derived from animals carries the risk of transmission of infectious diseases. Another application of collagen relates to the scaffold involving the cytokine and the signal peptide inside that may be useful in the fields of regeneration medicine and immunological therapy. Collagen may have potential for this use, but it has limitations due to the elasticity and the size of the mesh. With a small change in the sequence of the oligopeptide as the fundamental unit, a biomaterial with functions different from the original material would be realized at least in principle.

B. Nanoparticles and Nano-Sized Spaces

Nanotechnology as defined above can provide the materials, concepts, and unit processing to other fields such as information technology, electronics, and biomedical engineering. These fields can also provide materials, concepts, and processing techniques to the nanotechnology community. For example, if we think about the method of setting nanoparticles on a plane for the purpose of making a memory device and a sensor from a quantum dot, different kinds of answers could be provided. One answer would be using a protein known as chaperonin that holds a nano-sized space inside the particle. In this case, the unit process for setting the nanoparticles on a plane can be realized by the biomaterial holding the space inside.

Several other applications of the ability of a nanoparticle to hold the space inside have been developed. One is a nanoreactor for the purpose of accelerating a chemical reaction efficiently. Other applications would be liposomes for the purpose of delivering drugs as described above, although other uses are possible, for example, a particle holds a drug in the space inside, then delivers the drug to the target organ and releases it there. The SMANCS technique involves embolizing the organ that contains a hepatic cell cancer by intruding the probe upstream of the hepatic artery and releasing the particle. Another system delivers the drug into the liver via a nanoparticle with the space inside.

The B-type hepatitis virus includes a surface protein that has an affinity with hepatic cells.[24] The protein expressed in a yeast cell will be localized on the cell membrane. The area on the cell membrane where the protein is localized will become unstable, and the result is that the nano-sized spheric particle is separated from the membrane. Because the surface of the particle contains the membrane of the original cell and the surface protein of the type B hepatitis virus, the particle does not cause hepatitis in the animal or human into which it is introduced. This particle is not delivered *in vivo* to organs other than the liver; this may be verified by using the particle with a fluorescent dye. Progress in developing drug delivery systems will be made based on the idea of using a particle carrying a protein or peptide that demonstrates an affinity with a specific organ or individual cell within an organ.

C. Quantum Dot (Semiconductor Nanoparticle)

A quantum dot is a nanometer-sized metal and/or silicon cluster that has a distinct property of generating fluorescent light. In 1962, R. Kubo[25] discovered the quantum dot effect with a nano-sized metal cluster through theoretical calculations of quantum mechanical equations. The bulk metal was known to have a small-sized band gap in its electron orbit. Kubo calculated the electron orbit of the planar metal (with one-dimensional restriction) and obtained a higher band gap than that of the bulk metal (without dimensional restriction). Further calculation of the electron orbit of the metal wire (with two-dimensional restriction) led him to obtain a much larger band gap. Finally, he obtained the largest band gap with the calculation of the quantum-sized metal cluster (quantum dot) illustrated in Figure 2.3. In 1993, the quantum dot effect was experimentally shown by establishing a method for making the nanometer-sized metal cluster particles by self-organization.[26]

A quantum dot generates fluorescent light, the wave length of which depends on the size of the particle by the quantum size effect described above (see Figure 2.4). The incoming light with a wave length smaller than that of fluorescent light can cause the emission of an electron of the particle. This method allows use of a

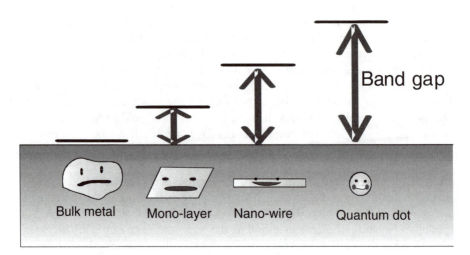

Figure 2.3   Band gap of metal cluster.

Quantum Size Effect

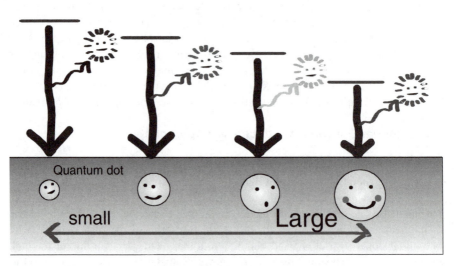

Figure 2.4   Quantum size effect and the fluorescence of nanoparticles.

much broader band of light for the emission than can be used with conventional organic compounds. Two photon emissions are also effective for the generation of fluorescent light. The quantum dot also demonstrates such a characteristic function as the light memory effect; the amount of the fluorescent light becomes higher after the emission and the memory can be erased by shining other light on it.

In the case of cadmium–selenium (Cd–Se) quantum dots, semiconductor nano-particles of Cd and Se are assembled in a single nanometer-sized reactor made by triocylylphosphine oxide at a high temperature (620 K). The Cd–Se semiconductor is covered with a shell such as ZnS in order to stabilize it, which results in generating core-shell-type semiconductor nanoparticles about 4 nm in diameter.

This nanoparticle dissolves in hydrophobic solvents but not in water. For use in biomedical research and engineering, hydrophilic surface treatment must be done to allow the particle to dissolve in water. After or during this process, biopolymer molecules such as proteins and nucleic acids can be conjugated with the quantum dot.[27,28] The applications of this specific supermolecule, for example, for detecting single molecules, imaging, and biological assays, have been reported in the biological and medical fields.[27,29–31]

The analysis of the mobility of cells and drugs inside the body using quantum dots has only started. After a cell has been marked with a quantum dot *in vitro* outside the body, the cell is introduced into the body. Especially inside blood vessels, cells marked with quantum dots are easily analyzed by the fluorescent activated cell sorter (FACS) system.[32]

Ruoslahti et al. reported on a quantum dot linked to a signal peptide delivered to the lung.[33] This study revealed the possible application of quantum dots conjugated with the drug to reach a targeted organ.

Cytotoxity is an important consideration for the application of quantum dots inside the human body. More suitable quantum dots or nanoparticles for the body have been developed based on materials such as silicon, platinum, titanium, and iron. The size of the particle is also important in order to allow it to pass through the urinary system. Some of the nanoparticles of quantum dots will meet these requirements. Most carriers for drug delivery systems including liposomes and block polymers are more than 10 nm diameter in size and cannot be eliminated from the body if they are not disassembled.

This chapter has presented an overview of self-assembled carriers for drug delivery systems. Although presently used carrier components differ from those of the 1980s, the sizes of the drug carrier components have reduced greatly — some are only a single nanometer in size. The size of a drug is estimated as a single nanometer. The size of a drug-conjugated quantum dot would not exceed 10 nm. Using a drug-conjugated quantum dot would allow us to follow drug mobility within the body, its organs, and even individual cells in real time. We could even control the target of the drug delivery, and this will provide a new development pathway for safer use of drugs.

V. SAFETY OF THE HUMAN BODY AND THE ENVIRONMENT

The safety of the human body and the environmental effects of the fabrication process are vital issues involved in both the treatment of diseases and the development of new single nanometer-sized drug carriers. A few cytotoxicity studies have been reported for newly developed functional nanoparticles such as water-soluble fullerenes[34–38] and quantum dots.[39] Minimal oral and dermal toxicity has been reported in animal studies of fullerenes[40] and an acute toxicity study performed after intravenous administration.[41] As for the quantum dot, Shiohara et al.[39] showed evidence of cell damage caused by the Cd–Se quantum dot with MTT assays and with a flow cytometry assay using propidium iodide staining. They also showed the existence of a threshold value for cytotoxicity. Hormone-disturbing agents are known to have no threshold concentrations for cytotoxicity; that means we have no way of using them safely on an industrial scale. The existence of a threshold value enables us to set maximum levels of concentration in drug delivery systems for use inside the human body and for release into the environment.

VI. CONCLUSION

This chapter reviewed technical developments in drug delivery systems based on self-assembled drug carriers used since the 1980s. This analysis was based on a chemical systems engineering concept by which the processes in living organisms, organs, and cells are reduced to chemical reactions. Later in this volume, Chapter 6 by Ineke Malsch and Chapter 7 by Emanuelle Schuler place these technical developments in a socioeconomic and nanotechnology research policy context.

REFERENCES

1. Matsumura Y and Maeda H. A new concept for macromolecular therapeutics in cancer chemotherapy. *Cancer Res.* 1986; 46: 6387–6392.
2. Oda T and H Maeda. Binding to and internalization by cultured cells of neocarzinostatin and enhancement of its actions by conjugation with lipophilic styrene–maleic acid copolymer. *Cancer Res.* 1987; 47: 3206–3211.
3. Konno T, Maeda H, Iwai K, Maki S, Tashiro S, Uchida M, and Miyauchi Y. Selective targeting of anti-cancer drug and simultaneous image enhancement in solid tumors by arterially administered lipid contrast medium. *Cancer.* 1984; 54: 2367–2374.
4. Ferruti P, Richardson S, and Duncan R. Poly (amidoamine)s as tailor-made soluble polymer carriers, in *Targeting of Drugs: Stealth Therapeutic Systems*, Gregoriadis G and McCormack B., Eds., Plenum Press, New York, 1988, pp. 207–224.
5. Liu M, Kono K, and Frechet JMJ. Water-soluble dendrimer–poly(ethylene glycol) star-like conjugates as potential drug carriers. *J. Polym. Sci. Polym. Chem.* 1999; 37: 3492–3507.
6. Landers JJ, Cao Z, Lee I, Piehler LT, Myc PP, Myc A, Hamouda T, and Baker JR, Jr. Prevention of influenza pneumonitis by sialic acid-conjugated dendritic polymers. *J. Infect. Dis.* 2002; 186: 1222–1230.
7. Fujita H, Ooya T, Kurisawa M, Mori H, Terano M, and Yui N. Thermally switchable polyrotaxane as a model of stimuli-responsive supramolecules for nanoscale devices. *Macromol. Rapid Commun.* 1996; 17: 509–515.
8. Hacein-Bey-Abina S, Von Kalle C, Schmidt M, McCormac MP, Wulffraat N, Leubouich P, Lim A, Osborne CS, Pawliuk R, Morillon E, Sorensen R, Forster A, Fraser P, Cohen JI, de Saint Basile G, Alexander I, Wintergerst U, Frebourg T, Aurias A, Stoppa-Lyonnet D, Romana S, Radford-Weiss I, Valensi F, Delabesse E, Macintyre E, Sigaux F, Soulier J, Leiva LE, Wissler M, Prinz C, Rabbitts TH, Le Deist F, Fischer A, and Cavazzana-Calvo M. LMO2-associated clonal T cell proliferation in two patients after gene therapy for SCID-X1. *Science.* 2003; 302: 415–419.

9. Kataoka K, Development of artificial viral vector for gene therapy: approach from polymer nanotechnology. *Biotherapy.* 2001; 15: 425–431.

10. Shah S, Sakthivel T, Toth I, Florence AT, and Wilderspin AF. DNA transfection and transfected cell viability using amphipathic asymmetric dendrimers. *Int. J. Pharm.* 2000; 208: 41–48.

11. Eguchi A, Kondoh T, Kosaka H, Suzuki T, Momota H, Masago A, Yoshida T, Aira H, Ishii-Watanabe AH, Okabe J, Hu J, Miura N, Ueda S, Suzuki Y, Taki T, Hayakawa T, and Nakanishi M. Identification and characterization of cell lines with a defect in a post-adsorption stage of Sendai virus-mediated membrane fusion. *J. Biol. Chem.* 2000; 275: 17549–17555.

12. Ben-Jacob E, Schochet O, Tenenbaum A, Cohen I, Czirók A, and Vicsek T. Generic modelling of cooperative growth patterns in bacterial colonies. *Nature.* 1994; 368: 46–49.

13. Eiha N, Hanaki K, Naenosono S, Yamamoto K, and Yamaguchi Y. Quantitative analysis of the growth of *Bacillus circulans* on a slant agar medium. *Bioimages.* 2000; 8: 129–133.

14. Komoto A, Hanaki K, Maeno S, Wakano JY, Yamaguchi Y, and Yamamoto K. Growth dynamics of *Bacillus circulans* colony. *J. Theor. Biol.* 2003; 25: 91–97.

15. Eiha N, Omoto A, Saenosono S, Wkano JY, Yamamoto K, and Yamaguchi Y. The mode transition of the bacterial colony. *Physica A.* 2002; 313: 609–624.

16. Zhang S. Building from the bottom up. *Mater Today.* 2003; 6: 20–27.

17. Samatey FA, Imada K, Nagashima S, Vonderviszt F, Kumasaka T, Yamamoto M, and Namba K. Structure of the bacterial flagellar protofilament and implications for a switch for supercoiling. *Nature.* 2001; 410: 321–322.

18. Vauthey S, Santoso S, Gong H, Watson N, and Zhang S. Molecular self-assembly of surfactant-like peptides to form nanotubes and nanovesicles. *Proc. Natl. Acad. Sci. USA.* 2002; 99: 5355–5360.

19. von Maltzahn G, Vauthey S, Santoso S, and Zhang S. Positively charged surfactant-like peptides self-assemble into nanostructures. *Langmuir.* 2003; 19, 4332–4337.

20. Santoso S, Hwang W, Hartman H, and Zhang S. Self-assembly of surfactant-like peptides with variable glycine tails to form nanotubes and nanovesicles. *NanoLetters.* 2002; 2: 687–691.

21. Zhang S, Marini D, Hwang W, and Santoso S. Designing nanobiological materials through self-assembly of peptides and proteins. *Curr. Opin. Chem. Biol.* 2002; 6: 865–871.

22. Hwang W, Marini D, Kamm R, and Zhang S. Supramolecular structure of helical ribbons self-assembled from a beta-sheet peptide. *J. Chem. Phys.* 2003; 118: 389–397.

23. Semino CE, Merok JR, Crane G, Panagiotakos G, and Zhang, S. A three-dimensional peptide scaffold culture system for enhanced hepatic stem cell differentiation. *Differentiation.* 2003; 71: 262–270.

24. Yamada T, Iwasaki Y, Tada H, Iwabuki H, Chuah MKL, VandenDriessche T, Fukuda H, Kondo A, Ueda M, Seno K, Fukuda H, Kondo A, Ueda M, Seno M, Tanizawa S, and Kuroda S. Nanoparticles for the delivery of genes and drugs to human hepatocytes. *Nat. Biotechnol.* 2003; 21: 885–890.

25. Kubo R. Electronic Properties of Metallic Fine Particles. *J. Phys. Soc. Jpn.* 1962; 17: 975.

26. Murray CB, Norris DJ, and Bawendi MG. Synthesis and characterization of nearly monodisperse CdE (E = S, Se, Te) semiconductor nanocrystallites. *J. Am. Chem Soc.* 1993; 115: 8706–8715.

27. Chan WCW and Nie S. Quantum dot bioconjugates for ultrasensitive nonisotopic detection. *Science*. 1998; 281: 2016–2018,

28. Hanaki K, Momo A, Oku T, Aomoto A, Naenosono S, Yamaguchi Y, and Yamamoto K. Semiconductor quantum dot/alubumin complex is a long-life, highly photostable endosome marker. *B.B.R.C.* 2003; 302: 496–501.

29. Han MY, Gao X, Su JZ, and Nie S. Quantum dot-tagged microbeads for multiplexed optical coding of biomolecules. *Nat. Biotechnol.* 2001; 19: 631–635.

30. Nie S and Emory SR. Probing single molecules and single nanoparticles by surface-enhanced Raman scattering. *Science*. 1997; 275: 1102–1106.

31. Nie S, Chiu DT, and Zare RN. Probing individual molecules with confocal fluorescence microscopy. *Science*. 1994; 266: 1018–1021.

32. Hoshino K, Hanaki A, Suzuki K, and Yamamoto K. Applications of T lymphoma labeled with fluorescent quantum dots to cell tracing markers in mouse body. *B.B.R.C.* 2004; 314: 46–53.

33. Akerman ME, Chan WCW, Laakkonen P, Bhatia SN, and Ruoslahti E. Nanocrystal targeting *in vivo*. *Proc. Natl. Acad. Sci. USA*. 2002; 99: 12617–12621.

34. Nakajima N, Nisshi C, Li FM, and Ikada Y. Photo-induced cytotoxicity of water-soluble fullerene, *Fullerene Sci. Technol.* 1996; 4: 1–19.

35. Sakai A, Yamakoshi Y, and Miyata N. Visible light irradiation of ^{60}fullerene causes killing and initiation of transformation in BALB/3T3 cells. *Fullerene Sci. Technol.* 1996; 7: 743–756.

36. Yang XL, Fan CH, and Zhu HS. Photo-induced cytotoxicity of malonic acid C-60 fullerene derivatives and its mechanism. *Toxicol. In Vitro*. 2002; 16: 41–46.

37. Chen HH, Yu C, Ueng, TH, Liang, CT, Chen B, Hong CC, Chiang, LY. Renal effects of water-soluble polyarylsulfonated C-60 in rats in an acute toxicity study. *Fullerene Sci. Technol.* 1996; 5: 1387–1396.

38. Chen HH, Yu C, Ueng, TH, Chen S, Chen BJ, Huang, KJ, Chaing, LY. Acute and subacute toxicity study of water-soluble polyarylsulfonated C60 in rats. *Toxicol. Pathol.* 1998; 26:1 143–151.

39. Shiohara A, Hoshino A, Hanaki K, Suzuki K, and Yamamoto K. On the cytotoxicity caused by quantum dots. *Microbiol. Immunol.* 2004; 48: 669–675.

40. Moriguchi T, Yano K, Hokari S, and Sonoda M. Effect of repeated application of C-60 combined with UVA radiation onto hairless mouse back skin. *Fullerene Sci. Technol.* 1999; 7: 195–209.

41. Rajagoparan P, Wudl F, Schinazi RF, and Boudinot FD. Pharmacokinetics of a water-soluble fullerene in rats. *Antimicrob. Agents Chemother.* 1996; 40: 2262–2265.

CHAPTER 3

Implants and Prostheses

Jeroen J. J. P. van den Beucken, X. Frank Walboomers, and John A. Jansen

CONTENTS

I. INTRODUCTION

The importance of prostheses and implants in medicine is growing. Due to increasing life expectancy, mankind will need a growing number of such synthetic devices to overcome the problems associated with deteriorating or failing body parts. Examples of implants are orthopedic joint prostheses, cardiovascular devices, dental implants, and others. An implant does not have to be located completely inside the body; skin-penetrating devices such as catheters for the infusion of fluids must be regarded as implants (see Table 3.1). Furthermore, the expected increase in the use of implants arises not only because certain devices are required because of medical reasons. The flourishing prosperity of the past few decades has meant that the use of implants for aesthetic reasons has become substantial.

Implants are made from biomaterials that have a common property: biocompatibility. Although biocompatibility is a difficult term to define, it is strongly related to the success of an implanted device in fulfilling its intended function.[1] This implies

Table 3.1   Applications of Synthetic and Modified Natural Materials in Reparative Medicine

Material	Application	Tissue Response
Titanium and its alloys	Joint prostheses, oral implants, fixation plates, pacemakers, heart valves	Inert
CaP ceramic	Joint prostheses, oral implants, bone replacement, middle ear replacement	Bioactive
Alumina	Joint prostheses, oral implants	Inert
Carbon	Heart valves	Inert
PTFE	Joint prostheses, tendon and ligament replacement, artificial blood vessels, heart valves	Inert
Poly(methylmethacrylate)	Eye lenses, bone cement	Tolerant
Poly(dimethylsiloxane)	Breast prostheses, catheters, facial reconstruction, tympanic tubes	Unknown
Poly(urethane)	Breast prostheses, artificial blood vessels, skin replacements	Inert
PLA	Bone fixation plates, bone screws	Inert
PGA	Sutures, tissue membranes	Inert

PTFE = poly(tetrafluoroethylene). PLA = poly(lactic) acid. PGA = poly(glycolic) acid.

that a biomaterial used for the manufacturing of a prosthesis or implant and subsequently classified as biocompatible cannot necessarily be used for the manufacturing of implants with different functions. For instance, biomaterials with properties that resist the adhesion of biomolecules and cells may be classified biocompatible when used for the production of cardiovascular devices. However, the classification may not be legitimate for the use of such a biomaterial to manufacture artificial joints.

Unfortunately, the classification of a biomaterial device as biocompatible does not necessarily imply its acceptance by its host. Most commonly, synthetic devices are recognized by their hosts as nonnatural and regarded as intrusions of foreign bodies.[2] For that reason, the possibility exists that although a biomaterial device is classified biocompatible, the coexistence of minor side effects of the implantation cannot be excluded. Although the potential side effects may not be detrimental to the functionality of the implanted device, they still can produce consequences that are not desirable.

Generally, the placement (for example, via surgical procedure) of and subsequent habituation (host reaction) to a synthetic device can be categorized as the establishment of a symbiosis of living and nonliving materials. This symbiosis is characterized by contacts of biological (e.g., biomolecules) and nonbiological compounds (e.g., molecules that constitute the biomaterial device). At this point, the utilization of biomedical nanotechnology may provide a contrivance to "smooth" the interactions between these molecules of dissimilar origin. For a logical comprehension about how biomedical nanotechnology might be able to achieve this, we have chosen to (1) first provide an overview of biomaterials used to date and their properties, (2) provide a general discussion of the biological processes that occur upon implantation of synthetic devices, and (3) focus on current and potential future applications of biomedical nanotechnology with respect to improving aspects of implantology.

II. BIOMATERIALS

A. Introduction

Biomaterials are substances used for the production of devices that interact with biological systems. This definition inherently suggests that biomaterials can be widely used. For example, biomaterials are used for cell cultures in laboratories, for the production of diagnostic devices, for extracorporeal apparatus (heart–lung machines), and many other applications. Nevertheless, the production of implants constitutes the main usage of biomaterials.

The use of biomaterials in medicine is not a novel concept. As early as 2000 years ago, dental implants made of gold or iron were used.[3] However, the practical use of implants in that era is not comparable to their present use. The increasing demands for safe and reliable implants have resulted in the evolution of biomaterials science as a distinct discipline. In addition to the somewhat old-fashioned biomaterials such as stainless steel, a high number of novel, mostly polymeric, biomaterials are now available and can be categorized as depicted in Figure 3.1.

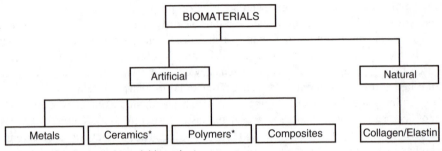

* Degradable and nondegradable variants

Figure 3.1   Classification of biomaterials.

B. Properties of Biomaterials

In order to function appropriately, biomaterials must possess properties that allow them to be used successfully for their intended applications. In view of this, it is logical to distinguish bulk properties from surface properties. Bulk properties (together with the design) determine the strength (mechanical) of an implant, whereas the surface properties are important in view of the eventual interactions of an implant with biological systems.

1. Bulk Properties

Bulk properties of materials are determined by the organization of the atoms of which the materials are built and the forces by which the atoms are kept together (interatomic forces). Three types of interatomic forces are known: ionic bonding, covalent bonding, and metallic bonding.[4]

The mechanical properties of a biomaterial must be adjusted to its intended function; otherwise the implant is likely to fail. For example, if a device intended for the fixation of a bone fracture lacks the required strength, it may break, making the device unsuitable for this function. Hence, the intrinsic properties of biomaterials may be appropriate for a certain application, and play a role in the failure of an implant made from the same material and used for another application. Regarding specific requirements from a mechanical view, three intrinsic properties of materials are especially important: elastic modulus, yield stress, and ultimate stress.[5] Together, these three parameters determine the stiffness, deformability, and strength of a material.

Another important bulk property of a biomaterial is fatigue — the "process by which structures fail as a result of cyclic stresses that may be much less than the ultimate tensile stress."[4] Such cyclic stresses are common at many locations in the human body, such as in a pumping heart (artificial heart valves), in the mouth (tooth prostheses), and at the connections of limbs (artificial hips).

2. Surface Properties

In addition to intrinsic (bulk) properties of biomaterials, surface properties are also important to the success of an implant. Since the interaction of a synthetic device and biological system takes place at the biomaterial–tissue interface, it is evident that the surface properties of a biomaterial are pivotal for the regulation of implant integration. Several factors determine the surface characteristics of a material, including composition, roughness, release of ions, charge, and energy.[6] Obviously, the interaction of biological constituents with the surface of a biomaterial device must not cause any detrimental effects to the surrounding viable cells, tissues, and organs. For that reason, the surface molecules of a biomaterial should not be toxic, carcinogenic, pyrogenic, cytotoxic, or antigenic to living cells in any way. If materials containing one or more of these characteristics are excluded, the reactions of body tissues to an implant surface still depend on the surface properties of the biomaterial in question.

One commonly assessed property for biomaterial surface characterization is surface energy. As will be discussed later, this parameter may be an important factor in the establishment of cell adhesion to biomaterial surfaces. However, other factors that are probably important in monitoring the adhesion of cells to biomaterial surfaces are cell type and the presence of adhesive proteins.[7] Interaction of biological systems with biomaterial surfaces can be desirable to enhance the integration of the biomaterial in the body. Additionally, the generation of noninteracting surfaces can be another aim in designing implants. In synthetic vascular grafts, for example, the deposition of biological material (biofouling or bioadhesion) is undesirable because it may lead to occlusion of blood vessels. Furthermore, control over bioadhesion may eventually result in the generation of biomaterial surfaces that encourage adhesion of host cells but discourage adhesion of infectious bacteria — a common cause of implant failure.[8-10]

Processes of a biological nature can affect the integrity of biomaterials. Upon implantation, a biomaterial is subject to interactions with the constituents of the biological environment and may be subject to biodegradation, a process in which components of the biological environment (or host) attack the biomaterial. Metals are inherently susceptible to corrosion — an electrochemical process in which oxidative and reductive reactions take place.[11] Due to such reactions, the integrity of a metal may be affected by the formation of metal ions from the solid metal, resulting in degradation. For ceramics, the extent to which the biological environment is capable of degradation is dependent on the chemical structure of the biomaterial. Increased knowledge in biomaterials science has resulted in the production of both polymeric and ceramic biomaterials whose degradation rates can be controlled. This has presented opportunities to generate and use biomaterial devices that in time can be replaced by native tissues.

C. Biomaterials Science: A Multidisciplinary Field

The success of an implant depends on a wide range of parameters that originate from many disciplines. The fabrication of implants involves know-how in the fields

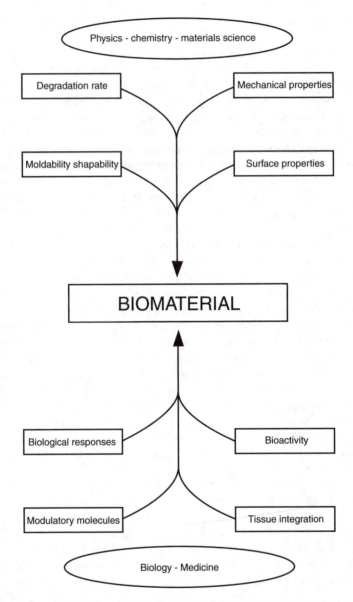

Figure 3.2   The multidisciplinary field of biomaterials science involves contributions of many
                scientific disciplines.

of materials science, physics, chemistry, and others. The *in vitro* and *in vivo* testing
and eventual application of implants require biological and medical knowledge. As
illustrated in Figure 3.2, several factors evolving from these divergent disciplinary
fields exert their impacts on the success of an implant. For that reason, collaboration
among researchers working in these fields is necessary for the directed expansion
of knowledge of biomaterials science.

III. BIOLOGICAL PROCESSES

A. Wound Healing Processes

The introduction of an implant into a living organism is commonly associated with surgical intervention. As noted earlier, an implant can be regarded as the intrusion of a foreign body that incontrovertibly initiates a response of the body toward the intruder through specific reactions arising from both the introduced material and the tissue damage or injury caused by the surgical procedure. In general, the response consists of wound healing processes that have the ultimate aim of healing the affected tissues, preferably without permanent damage. While this phenomenon, also known as *restitutio ad integrum*, is feasible for human and animal fetuses, wound healing produces scar formation in adult humans and higher vertebrates.[12]

Evidently, fibroproliferative response rather than tissue regeneration controls the repair of soft tissues. At this point, a striking difference exists between healing processes in soft tissues such as skin and hard tissues such as bone. While soft tissue healing is reparative, regenerative healing in hard tissues occurs after wounding.[13] This means that the formation of scar tissue is absent in healing of hard tissues. Although its appearance is radiographically different, healed hard tissue will eventually return to its pre-injury state[14] and possess the same or even improved characteristics compared with the original tissue.

The biology behind the processes of soft and hard tissue healing has been studied extensively, and several excellent reviews on this subject have been published.[13,15–22] Although not all their mysteries have been revealed yet, it is already evident that the healing processes involve a tight and regulated collaboration of specific cell types and their signaling products.[23,24]

The wound healing processes preceded by thrombus formation are divided roughly into three overlapping phases: (1) inflammation, (2) repair, and (3) tissue remodeling.[13,15,16]

1. Thrombus Formation

Prior to the initiation of the healing processes, fluids containing blood constituents surround the newly implanted biomaterial. The fluids originate from the disruption and increased permeability of blood vessels and the subsequent extravasation of blood constituents, all of which are consequences of tissue injury. Through changes in the environment, several components of the blood including platelets and the surrounding or even adjacent tissues become activated, thereby initiating the blood coagulation cascade.[19] The activation of platelets results in increasing adhesiveness. This enables platelets to aggregate and form a plug to close perforations of damaged vessels and thus limit blood loss. The coagulation cascade also involves polymerization of fibrin. Activated platelets and strands of polymerized fibrin together form a fibrous clot that serves as a matrix for subsequent migration of a variety of cells into the area of injury. The recruitment of cells to the area of injury is at least partially orchestrated through the release of certain biologically active substances by platelets[25] and endothelial cells.[26] The migratory cells include those that are important to the inflammatory response, the formation of new tissue, and the tissue remodeling processes.

2. Inflammatory Phase

Inflammation is a physiological response of tissue resulting from detrimental physical, chemical, or immunological stimuli or from infection.[27] The inflammatory response is initiated as a reaction to the release of vasodilators, chemoattractants, and other mediators, including platelet-derived growth factor (PDGF) and tumor growth factor-beta (TGF-β) by platelets[28] and activation of the complement cascade within the coagulating fluid surrounding the implanted biomaterial.[16] The release of these substances is responsible for the recruitment of inflammatory and other cells (chemotaxis), the development of new blood vessels (angiogenesis), and overall cell regulation at the site of injury. The response consists of nonspecific defense mechanisms carried out by cells and noncellular components of the circulating blood (granulocytes, monocytes, and the complement system) as well as resident inflammatory cells (macrophages and mast cells) that collectively try to eliminate intruders. If necessary, specific immune responses such as the production of antibodies by B lymphocytes and/or the activation of cytotoxic T lymphocytes can be initiated.

3. Reparative Phase

The formation of new tissue requires the activation and/or proliferation of distinct cell types, resulting in the replacement of lost or damaged tissue. In soft tissue healing, extremely important cells related to new tissue formation are fibroblasts and endothelial cells that are capable of new extracellular matrix formation and angiogenesis, respectively. The provisional extracellular matrix is important as a scaffold for the migration of cells into the damaged area. Additionally, the extracellular matrix and its components contain signals for the differentiation and stimulation of cells, mainly via receptor–ligand interactions. Through the development of new blood vessels, nutrients and oxygen become available for proliferating cells that replace the tissue at the damaged area.

In hard tissue healing, the process of ossification (bone formation) is important and two mechanisms assure new bone formation: intramembranous and endochondral ossification.[13] Intramembranous ossification is carried out by osteoprogenitor cells present in the cambium layer of the periosteum. Endochondral ossification occurs at and overlies the defect site and undifferentiated mesenchymal cells attracted from tissues surrounding the defect (e.g., soft tissues and periosteum) become committed cartilage-producing cells[21] under influence of local production and release of mediators, including growth factors. The mineralization of the cartilage tissue leads to bone formation.

4. Tissue Remodeling

Tissue remodeling involves the transition of newly formed, immature tissue into mature tissue. In contradistinction to both the inflammatory and reparative phases, the remodeling phase may last for several years. The general process of soft tissue remodeling involves rapid synthesis and degradation of connective tissue proteins.[29] The degradation of these extracellular matrix (ECM) proteins is accomplished through the actions of matrix metalloproteinases (MMPs).[30] The common outcome of soft tissue remodeling is scar formation, which results mainly from an imbalance between the stimulation of collagen synthesis and degradation of extracellular collagen.

Remodeling in hard tissues involves bone resorption by osteoclasts, followed by the synthesis of new bone matrix and its mineralization by osteoblasts. The remodeling process in hard tissues is subject to mechanical forces acting upon it (Wolff's law[31]). In contrast with soft tissue remodeling, hard tissue remodeling is devoid of scarring. Furthermore, healed hard tissue is able to resume its original configuration.

B. Macrophages

Several cell types are involved in the biological processes that occur after the implantation of a biomaterial. The interplay among these cells is extremely important because inadequate cellular responses could directly or indirectly impede the functionality of the implanted device. Cells respond to stimuli mostly via receptors on their surfaces. Via these receptors, cells can recognize a large variety of ligands including soluble mediators secreted by other cells (cytokines), molecules present on the surfaces of adjacent cells, and distinct patterns in molecules of ECM proteins.

Due to their early appearance at an implantation site, their longevity, and the large number of cytokines they can produce and secrete, macrophages are generally considered the most important cell type in the vicinity of a newly implanted device.[32] Macrophages perform multiple functions at a site of implantation ranging from phagocytosis of cell debris and potential pathogens via initiation of an inflammatory reaction to orchestration of the processes necessary to heal the damaged tissue resulting from the surgical procedure. In summary, the macrophages at an implantation site govern the magnitude and duration of all phases and subphases of the wound healing process by means of the versatility in the mediators they secrete that control the responses and functions of many other cell types.

C. Biomaterial Interface Processes

Although an implant is subject to cellular biological processes upon introduction, as described above, the initial contact of implant and host relies on noncellular interactions. A newly introduced implant is surrounded by an aqueous liquid. The water molecules in the direct vicinity of an implant can substantially alter the appearance of the biomaterial surface for the biological environment.[33] The abundance of water molecules within this liquid means that water is the primary molecule involved in the first series of interactions of a biomaterial surface interface with its *in vivo* surroundings.

An important parameter is the free energy of a biomaterial surface reflected by its water wettability. Biomaterial surfaces are often categorized as hydrophobic or hydrophilic. A related parameter of biomaterial surfaces is cell adhesion. Although some authors assert the existence of a correlation between surface free energy and cell adhesion,[34,35] others impugn this correlation[36] or even postulate the inverse.[37] The water molecules in the direct vicinity of the biomaterial surface will form a water monolayer or bilayer in which the arrangement of the water molecules depends on the surface properties at the atomic scale and completely differs from that of liquid water.[33]

Subsequent to interface interactions with water molecules, a biomaterial surface will first encounter ions and then the proteins present within the surrounding liquid. In the monolayer or bilayer of water molecules, natural ions (e.g., Na^+ and Cl^-) are incorporated as hydrated ions.[38] The surface properties of the biomaterial determine the type, amount, and conformational state of the adsorbed proteins.[39,40] Thus, the spectrum of adsorbed proteins will not necessarily reflect the amounts and ratios of the proteins within the surrounding liquid.[41,42] Additionally, denaturation of the adsorbed proteins may occur. As a result, biologically important sites may become inaccessible or nonfunctional, limiting interactions with counter-receptors present on cellular membranes.

Finally, living cells will become involved. The presence of a wide variety of membrane-bound receptors on the surfaces of cells enables them to adhere to adsorbed proteins on the biomaterial surface. Because the interaction of cells with the biomaterial surface does not rely on direct contact between cells and biomaterial, but merely on an indirect interaction mediated by adsorbed proteins, it has been suggested that the biomaterial is not what causes unwanted responses.[43] The non-specific layer of proteins adsorbed on the biomaterial surface immediately after implantation is recognized by the host as a foreign or unnatural material. This assumption seems plausible because such an adsorbed mixture of proteins with random orientations and conformational states presents a divergence from natural, intentionally arranged protein layers.

D. Foreign Body Reaction

The cumulative effects of all separate contributive processes that occur at the biomaterial interface result in one of the following outcomes of implantation: (1) integration, (2) extrusion, (3) resorption, or (4) encapsulation. Although integration of the biomaterial device is the most favorable outcome, the number of cases in which true biointegration is achieved is limited.[44] Most frequently, true biointegration occurs after implantation of compatible biomaterials such as titanium coated with hydroxyapatite (HA) into bone tissue.[45,46] Implantation of biomaterials into soft tissues usually results in one of the other three outcomes.

Extrusion occurs when an implanted device is in direct contact with epithelial tissue. The epithelium will form a pocket continuous with the adjacent epithelial membrane that subsequently dissipates the implant. In the case of external epithelium, the implant will be externalized from the host. Resorption of the implant can occur when an implant is made of degradable material. After complete resorption, only a collapsed scar will remain at the implantation site. In most cases, implanted biomaterials in soft tissues become encapsulated by a process known as the foreign body reaction[2,47] (Figure 3.3). The capsule commonly consists of a relatively hypo-cellular membrane with a high collagen content.[48] Adjacent to this collagenous

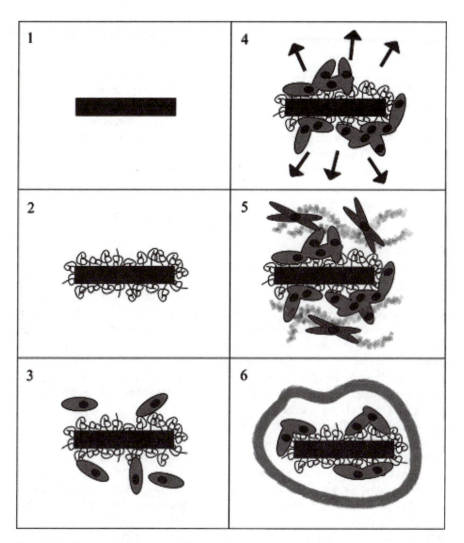

Figure 3.3   Foreign body reaction. The introduction of an implant (1) into a receptor leads to the adsorption of proteins in all possible configurations on the surface (2). Subsequently, cells (including macrophages) will attach to the implant surface via cell surface receptors that recognize corresponding ligands in the adsorbed protein layer (3). Attached cells secrete a wide variety of signal molecules that influence the behavior of perceptive cells (4) that become activated and start to produce extracellular matrix (5). Finally, the implant becomes enclosed in a fibrous capsule that isolates the implant from the body (6).

membrane, a layer of myofibroblasts is occasionally observed. Furthermore, foreign body giant cells (FBGCs; fused macrophages) are frequently observed in the space between implant and capsule.[49,50]

In general, the organization of cells and matrix surrounding an implant is built up in such a way that a barrier between the foreign material and the body is created, and this structure more or less isolates the implant from the body. The capsule, including FBGCs, surrounding the implant may persist for the lifetime of the implant. However, it is not yet clear whether FBGCs present at the biomaterial surface remain activated during the lifetime of the implant or become quiescent.[47] Since encapsulated implants can perform their functions for many years, the isolation of an implant in a collagenous capsule is not necessarily an unwanted phenomenon. It may even help the body live in symbiosis with a synthetic device, although the presence of genuine symbiosis in this respect may be arguable. Unfortunately, the presence of myofibroblasts within the capsule may lead to contraction and thus cause pain and/or implant failure. Furthermore, the formation of a capsule associated or not associated with wearing of the biomaterial may result in loosening.

IV. NANOTECHNOLOGY IN IMPLANTOLOGY

A. Introduction

From the previous descriptions of biomaterial properties and interfacial biological processes, it is evident that the placement of an implant into a living organism causes specific reactions of the biological environment. The biomolecules and cells on the one hand and the intrinsic properties of the biomaterials on the other determine the biocompatibility and longevity of synthetic devices. Since the interaction of biomolecules and cells with the biomaterial surface is a vital element in evaluating the suitability of a biomaterial for its intended function, it is not necessary to note that every attempt to avoid undesired responses and/or enhance desired responses to implants is of utmost interest.

In many disciplines including biomaterial science, miniaturization has been a topic of interest for several years[51] and led to the evolution of microtechnology techniques[52,53] that allow the creation of features with microscale dimensions on biomaterial surfaces. Further expansion of many of these techniques, development of novel techniques, and focusing on medical applications resulted in expansion of the field of biomedical nanotechnology[54] dealing with dimensions 1000-fold smaller than previously possible. In general, the emerging field of nanotechnology aims to increase control over material structures of nanoscale size in at least one dimension (x, y, or z).[55] As already shown, microscale features can exert control over cellular behavior,[56-59] and recent improvements in the field of nanotechnology may yield powerful additional tools to increase control over reactions of the biological environment to submicron cues in the direct vicinity of a biomaterial device.[60]

The general difference between microtechnology and nanotechnology is the size of the created microscale or nanoscale structures. Generation of nanoscale structures can be based on the miniaturization of higher scale structures (top-down) or on the assembly of nanoscale structures from ultimately small structures (bottom-up). The convergence of top-down and bottom-up strategies to create features with nanoscale dimensions via the collaboration of many scientific disciplines, for example, chemistry, physics, biology, and medicine, makes it possible to produce materials that resemble natural surroundings for biological entities.

The three-dimensional organizations of structures surrounding cells *in vivo* influence most cellular processes, e.g., adhesion, migration, growth, differentiation, secretion, and gene expression. The majority of such structures such as ECM components and membrane-bound receptors on cells encompass dimensions down to nanoscale size. The organization of cells and ECM proteins has been hypothesized to be of importance in controlling cellular behavior, and this was shown in an elegant experiment using multigrooves (a combination of microgrooves and macrogrooves).[61] The experiment demonstrated that control over both cellular orientation and ECM orientation is feasible. Consequently, it was suggested that multigrooves may allow the production of three-dimensional ECM *in vivo*.

The introduction of nanodimensional structures on the surface of a biomaterial is possible by means of present nanotechnology, and such structures may influence biological reactions to implants and prostheses. Although distinct from natural nanostructures, synthetic nanostructures may be able to influence cellular responses to biomaterial implantation. Because nanotechnology is still in its infancy, future developments could expand the efficacy and thus the importance of creating nanostructures on biomaterial surfaces. A number of nanotechnology-based methods to modify biomaterial surfaces are described below and the effects of such nanotechnologically modified biomaterial surfaces on cell behavior will be discussed.

B. Current Nanofabrication Methods

The production of nanostructures on biomaterial surfaces is an emerging field of technology that may involve utilization of many techniques. Several, but certainly not all, methods for the fabrication of biomaterial surfaces with nanoscale topological or chemical cues are listed in Table 3.2. Their principles are described below. In general, nanotechnological modifications of biomaterial surfaces can be categorized into those that alter a surface topographically and those that introduce nanoscale chemical molecules (or groups) on a surface. The techniques described below, however, do not necessarily restrain themselves to one of these types of modifications. Many techniques can serve multiple purposes, for example, using some kind

Table 3.2 Available Methods for Nanofabrication of Biomaterial Surfaces

Type of System	Materials	Resolution
Lithography	Silica, silicon, silicon nitride, silicon carbide	x, y, and z to 10 nm
Colloidal resist	Silica, silicon, silicon nitride, silicon carbide	x, y, and z to 5 nm
Self-organizing or self-assembling	Polymer demixing, self-assembling particles and monolayers, other self-assembling systems	In 10-nm range
Soft lithography	Any fairly large molecule	x and y to 200 nm, z to one monolayer
Biomimicry	Many	Actual native dimensions

Source: Partly adapted from Curtis, A. and Wilkinson, C. *Trends Biotechnol*, 19, 97–101, 2001.[130]

of mask may involve either topography or chemistry. Topography can be further specified as having either texture or roughness. The difference between texture and roughness is determined by the regularity of the topographical cues. While texture is characterized by an organized regularity in topography, roughness encompasses a random topography.[62]

1. Lithography

Figure 3.4 depicts the basic principles of photolithography. Lithography is a technique by which a material is coated with a film prior to the creation of desired features. The film is usually a polymer that is sensitive to a particular type of energy applied. Polymers sensitive to light or to electrons can be used. Depending on the sensitivity of the polymer (also called the resist), lithographical techniques are categorized as photolithography (light-sensitive resist) or electron beam lithography (electron-sensitive resist).

The irradiation of a specific pattern in a sensitive polymer modifies the polymer properties in that area. A subsequent dissolution step removes the affected sensitive polymer, leaving a specific pattern of sensitive polymer at the surface of a biomaterial. Photolithography commonly employs a mask to allow control over the irradiation of the resist, whereas in electron beam lithography, the beams of electrons can be focused at and maneuvered to the desired positions to gain control over the

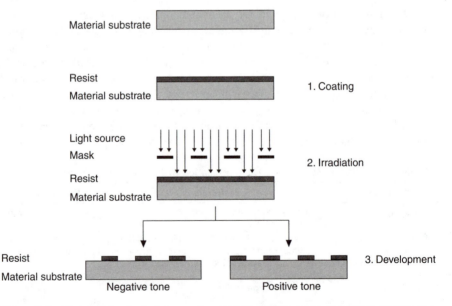

Figure 3.4 Photolithography techniques. In conventional lithography, a resist is coated on a material substrate and the resist is subsequently irradiated through a mask, creating a pattern corresponding to the mask in the resist. Development of the resist will result in a positive or negative tone on the material surface that can be used for coating or etching techniques.

irradiated zone. Two types of further modification of the surface from which the polymer has been removed can be applied: (1) etching and (2) film deposition. Etching allows pits, grooves, and other topographies of controlled shape and size to be created. On the other hand, the deposition of a thin film basically relies on coating the exposed area with a desired solution, from which the solvent evaporates or in which the particles (molecules) organize themselves in a specific conformation (self-assembly). The selectivity of and the precision by which the energy used to irradiate the sensitive polymer is applied determine the range of the dimensions of patterns that can be created. Generally, the resolution of conventional photolithography is 300 nm, whereas lithography features down to 10 nm in size can be created with electron beam lithography.[38a]

2. *Colloidal Resists*

In addition to masks (as in photolithography) or precision maneuvers of an electron beam (as in electron beam lithography) the application of colloidal particles is possible (Figure 3.5). Colloidal particles of different materials and sizes down to 5 nm can be produced and subsequently dispersed over a biomaterial surface. The distribution (e.g., density) of the particles on a surface can be controlled by the salinity[38b] and acidity (pH) of the solution. Subsequently, the adsorbed particles can be used as a template for patterning the underlying surface. In a technique similar to photolithography and electron beam lithography, the space not covered by colloidal particles can be etched or a thin film can be deposited. After removal of the colloidal particles, a patterned surface remains. Using colloidal resist techniques followed by etching or thin film

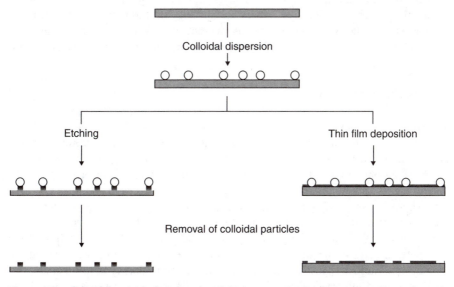

Figure 3.5   Colloidal resist techniques. A colloidal suspension is dispersed on the surface of a material. Subsequent etching or coating, followed by removal of the colloidal particles, results in a pattern on the material surface.

deposition and subsequent removal of the colloidal particles, the variations in the pattern are related to particle size and spatial distribution.

3. Self-Assembly Systems

Self-assembly is a common phenomenon in nature. It is described as a "spontaneous association of numerous individual entities into a coherent organization and well-defined structures to maximize the benefit of the individual without external instruction."[63] If this phenomenon is downscaled to smaller entities, molecular self-assembly results. Molecules organize spontaneously into structurally well defined and rather stable arrangements via noncovalent interactions under equilibrium conditions.[64]

The formation of a cell membrane from single phospholipid moieties is a good example of naturally occurring self-assembly at the molecular scale. It becomes apparent from this example that self-assembly allows the formation of stable structures, whereas single noncovalent interactions may be somewhat weak and collective interactions can be more than sufficient to create very stable structures and materials. The establishment of a self-assembling system relies on chemical complementarity and structural compatibility. Therefore, a vital prerequisite for self-assembly is the use of molecules of correct size and orientation (chirality). Monolayers of molecules with distinct properties exposed at the "new" surface can be generated and are designated self-assembling monolayers, or SAMs.

One common application of SAM technology is protein patterning. The generation of a self-assembled monolayer of molecules (e.g., alkylsilanes or alkane thiol molecules) into an organized layer[10] results in the possibility of effectively modulating the properties of the layer of free end groups. Via variation of unique reactive end groups in the SAM, homogeneous interactions (hydrophilic end group with hydrophilic protein and v.v.) with proteins can provide a mechanism of protein patterning.

4. Soft Lithography

Soft lithography (Figure 3.6) is a term collectively used for a group of lithographic techniques in which a patterned elastomer, usually poly(dimethylsiloxane) or PDMS, is used to generate or transfer this specific pattern via molding, stamping, or masking onto a biomaterial surface. Additionally, the PDMS inverse replica can be used as a "master" to generate positive replicas of the original template.

Microcontact printing is a soft lithographic technique using the contact of the relief pattern of the PDMS stamp with the biomaterial surface to generate a pattern on the latter. Prior to the moment of contact between the PDMS stamp and the biomaterial surface, the stamp is "inked" to create the pattern of the stamp at the biomaterial surface. Most commonly, microcontact printing is used together with SAMs on gold substrates. Patterns of specific SAMs can be created, after which the intrapattern space can be filled using another SAM. SAMs have, respective to their chemical properties, selective adsorption profiles for proteins. Selecting appropriate SAMs and designing them into a pattern can control protein adhesion. Such

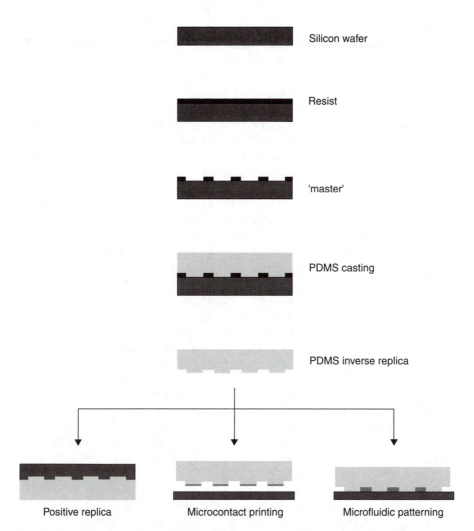

Silicon wafer

Resist

'master'

PDMS casting

PDMS inverse replica

Positive replica   Microcontact printing   Microfluidic patterning

Figure 3.6   Soft lithography techniques. Using conventional lithography techniques, a master
is prepared, onto which PDMS is cast. The PDMS inverse replica can subsequently
be used to create patterns (via etching, coating, etc.) on material surfaces via
techniques like casting, microcontact printing, and microfluidic patterning.

patterned, protein-containing surfaces can serve as ligands for cell receptors, thus
providing the opportunity for directed cell attachment.[65] In addition to indirect
protein immobilization through SAMs, direct patterning of proteins using microcontact printing is also possible.

Microfluidic patterning is a technique using the network of microchannels created during contact of the PDMS stamp for the generation of patterns on a biomaterial
surface. Via these microchannels, fluids can be delivered to selected areas of a
substrate. In microcontact printing, the pattern is created at the sites of contact
between stamp and biomaterial. In contrast, in microfluidic patterning, the areas
where the stamp is not in contact with the biomaterial are responsible for the

patterning. Depending on the type of fluid used, several possibilities for creation of a pattern are feasible: (1) solidization of the fluid, (2) deposition of soluble constituents, or (3) removal of underlying material.

5. Biomimetic Approaches

A completely different approach regarding the modulation of implant surfaces is the use of biomimicry. Biomimetic approaches attempt to create an implant surface, which is not, or to a lesser extent, recognized as foreign by the host. Constituents of the natural cellular environment (i.e., ECM proteins) that often have nanoscale dimensions can be of help in creating biomimetic surfaces.

Under natural conditions, cellular functions are regulated via interactions of cells with their direct surroundings, and cells recognize specific components of their surroundings, including ECM components. For that reason, research has focused on mimicking such surroundings on biomaterial surfaces, both topographically and biologically. Much effort has been devoted to creating biomaterial surfaces that contain elements of native ECM proteins. Such proteins have been demonstrated to contain domains that can influence cell behavior. Receptors located on the surface of a cell can recognize such domains that can function as their counterparts (ligands; key–lock principle).[66] The interactions of the receptor family of integrins with such domains are particularly known for their impacts on cellular processes.[67,68] For example, adhesion of cells to specific domains of ECM proteins can be achieved via receptor-mediated interactions.

Additionally, receptor-mediated interactions can influence other cellular processes including proliferation, migration, morphological change, gene expression, and cell survival by intracellular signaling. The introduction of native ECM components onto the surfaces of biomaterials is an interesting modification method that can generate a biomaterial interface akin to a natural one (biomimicry) onto which cellular behavior can be influenced. An additional prospect of using ECM components for the generation of biomimetic surfaces involves the capacity of ECM components to strongly bind growth factors[69,70] that can further modulate cellular behavior, depending on the type of growth factor applied.

In general, three major methods exist for the immobilization of biomolecules such as proteins and peptides onto surfaces: (1) physical adsorption (e.g., via van der Waals or electrostatic interactions); (2) physical entrapment (use of a barrier); and (3) covalent attachment. In addition to these methods, more sophisticated techniques such as covalent linking to polymeric networks can be used to generate biomimetic surfaces containing elements of native ECM components.[71,72]

Although adsorption of entire proteins (e.g., fibronectin) is demonstrated to be effective in enhancing cellular attachment,[42] research has focused on the design of materials representing only parts of ECM proteins. Generally, these parts (or peptides) are based on the primary structure of the receptor-binding domain of an entire protein such as fibronectin or laminin. These peptides, whether linear or cyclized, can possess similar functionalities, for example, receptor specificity, binding affinity, and signaling of cell responses, compared to their native proteins.[73,74]

A major opportunity in using peptides instead of complete proteins is to target specific cellular interactions to a given peptide, while eliminating possible undesired responses of an intact protein. Furthermore, displaying short peptides appeared to enhance the availability and activity of receptor-binding domains as compared with displaying the entire native protein.[75] Presumably, the use of entire proteins is associated with many possible orientations and occasional sterical hindrance, resulting in a less effective display of the receptor-binding domains as compared to short peptides. Although several domains are known to be beneficial in the enhancement of cell binding to biomaterial surfaces,[76] peptides containing the arginine–glycine–aspartic acid (RGD) amino acid sequence are mostly used. This tripeptide is the cell-binding domain of fibronectin, and known to serve as a ligand for an integrin receptor ($\alpha_5\beta_1$) expressed on the surfaces of many cells and involved in many cellular processes, including adhesion, migration, assembly of ECM products, and signal transduction.[77]

The previously mentioned modifications of biomaterials involving elements of native ECM can be useful for enhancing tissue integration of implants in both soft and hard tissues.[78–80] However, since natural hard tissues comprise precipitated minerals, they are also used for creating biomimetic biomaterial surfaces. The most important inorganic constituents of biological hard tissues such as bones and teeth are calcium phosphates, and they are widely used as biomaterial surface coatings for bone implants. Furthermore, calcium phosphates are bioactive, which means that they allow dynamic interactions favoring bone formation with implant surroundings.[6,81] Many techniques have been developed to deposit calcium phosphates on biomaterial surfaces, including magnetron sputtering techniques,[82,83] plasma spraying techniques,[84] and the novel electrostatic spray deposition technique.[85] These techniques allow the generation of nanostructured calcium phosphate coatings with several potential phases of calcium phosphate.

6. DNA Coatings

Another example of nanoscale modifications on biomaterial surfaces deals with the generation of DNA-containing coatings for biomaterial purposes. The hypothesis is that DNA can have several advantages when used as a structural element, regardless of its genetic information. Vertebrate DNA, a natural polymeric material, is regarded as nonimmunogenic or slightly immunogenic,[86] unlike bacterial DNA, a potent stimulator of immune reactions.[87,88] This difference in immunostimulatory reaction is due to an abundance of unmethylated cysteine–phosphate–guanine (CpG) dinucleotides in bacterial DNA.[89] Additionally, DNA can be used as a drug delivery vehicle.

The structure of DNA allows its interaction with other molecules via mechanisms including groove binding and intercalation.[90–92] In view of this, the loading of DNA with molecules that elicit specific cellular responses (cytokines, growth factors, antibiotics, etc.) can deliver these signal molecules at an implantation site. A third application of DNA may be its use as a suitable bone deposition material. Since phosphate groups favor the deposition of calcium phosphate,[93,94] the high content of phosphate groups in DNA may also favor the deposition of calcium phosphates.

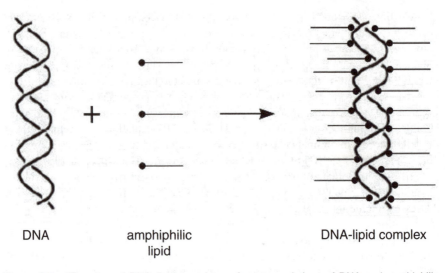

DNA   amphiphilic   DNA-lipid complex
lipid

Figure 3.7   Formation of DNA–lipid complexes. Aqueous solutions of DNA and amphiphilic lipids are mixed in an appropriate phosphate anion-to-amphiphilic lipid cation ratio. Formation of DNA–lipid complexes is accompanied by their precipitation in the aqueous (mixed) solution. Via subsequent wash steps and lyophilization, dry DNA–lipid complexes that are soluble in organic solvents are produced.

Finally, DNA–lipid complexes, depending on composition, may exert antibacterial activities.[95] Since infections are common problems associated with implantation procedures, a coating that possesses antibacterial activity may diminish the incidence of implantation-related infections.

The use of DNA as a nanocoating on a biomaterial surface, however, implies the necessity to circumvent certain properties of DNA, including its water solubility and easy degradation by nucleases. DNA can be complexed with amphiphilic lipids[96,97] (Figure 3.7) or cationic polyelectrolytes[98] (Figure 3.8). The structures generated by this process are stable through electrostatic interactions between anionic phosphate groups in the DNA and cationic groups in the amphiphilic lipid or polymer. The application of DNA coatings in implantology may lead ultimately to multifunctional coatings that can be applied at various sites in the body, evoke minimal immunologic reactions, and deliver biologically active substances to modulate cellular behavior.

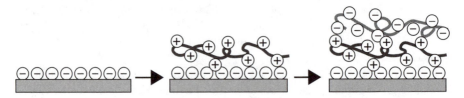

Figure 3.8   Formation of multilayered polyelectrolyte coatings. Polyanionic (e.g., DNA) and polycationic polymers can be used to generate multilayers based on electrostatic interactions between alternate layers. This technique allows a wide variation in the number of polyelectrolyte layers that form a multilayered coating and the types of polyelectrolytes.

C. Influence of Biomaterials with Nanostructures on Cell Behavior

This section describes the influence of nanostructured biomaterials on cell behavior based on a selection of recently published research work. Due to their recent development, nanotechnologically modified implants have not achieved clinical applications yet. Before clinical application is possible, *in vitro* and *in vivo* test models must demonstrate the benefits of nanotechnologically modified implants.

In our view, nanostructured biomaterials contain features that possess at least one dimension (x, y, or z) in the submicron (<1 μm) range. Although classification of nanotechnological methodologies is difficult, we have tried to generate an overview of cell behavior in relation to biomaterials with topographical, protein–peptide, and calcium phosphate nanostructures.

1. Topographical Nanostructures

The topography of biomaterial surfaces has been a major topic in biomaterials science in the past decade. Surface topography can be of great importance with respect to area enlargement. An increase in surface area may provide greater potential for tissue integration (mechanical interlocking). Excellent reviews on this topic[56,57,99–101] evidence a general consensus that topography indeed influences cell behavior. In the first studies that explored the effect of topography on cell behavior, microscale topographical cues were usually used. An enormous diversity of topographical cues was used: grooves, pits, ridges, cliffs, tunnels, steps, waves, wells, tubes, nodes, pillars, pores, spheres, and cylinders. Researchers used many different cell types in studies to examine the effects of microscale topographical cues on the behavior of primary isolated cells or immortalized cell lines including fibroblasts, macrophages, epithelial cells, leukocytes, neuronal cells, endothelial cells, and osteoblasts.

Although the reaction is dependent on cell type, cells react on contacting microscale topographies in a wide variety of manners including orientation, extension, movement, and activation [phosphorylation, actin polymerization, messenger ribonucleic acid (mRNA) expression, and phagocytic activity]. A phenomenon called contact guidance is observed when cells are cultured on microgrooved substrata (Figure 3.9); the cells align along the axes of the grooves. Control over cellular alignment (including the alignment of cell extensions) may be a pivotal factor in orchestrating cell morphology and orientation for the generation of nerve and other well-organized tissues.

Unfortunately, the precise biological effects of microscale topographies remain unclear — various research groups have obtained contradictory results. Parker et al.[102] found no favorable effects of surface texturing on capsule formation around subcutaneous implants, but *in vivo* studies by others indicated that grooved implant surfaces produced beneficial effects on tissues surrounding the implant.[99] In the latter studies, grooved topographies appeared to encourage tissue organization and showed a reduction in fibrous capsule formation as compared to smooth implant surfaces.

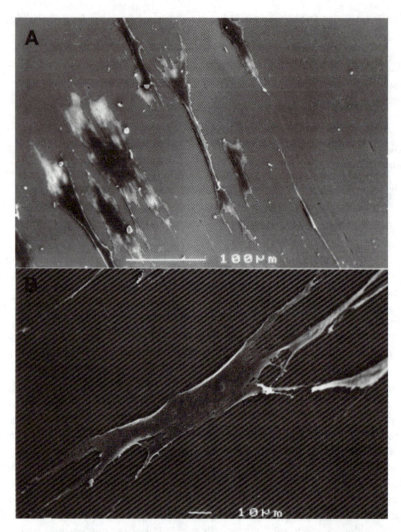

Figure 3.9   Contact guidance phenomena of rat dermal fibroblasts on microgrooved substrates. (A) Cells align themselves along the axes of the microgrooves (groove width = 1 μm, ridge width = 1 μm, and groove depth = 1 μm). (B) Higher magnification showing cell extensions along the axes of the microgrooves.

Although the technology for creating topographical cues on biomaterial surfaces with nanoscale dimensions is a novelty, several studies have demonstrated the substantial effect the cues can exert on cell behavior. The effect on cell behavior of grooves with nanoscale dimensions has been a subject of interest in many *in vitro* studies.[103–106] Regarding the effects of grooves with nanoscale dimensions, the findings generally indicate that cells become oriented and elongate along the surface grooves. Furthermore, the activity of cells on nanotopographically modified substrates is increased compared to activity on smooth control substrates. This is demonstrated by higher proliferation rates and enhanced spreading of fibroblasts,[107] increased phagocytic activity of macrophages,[106] and a notable up-regulation of gene expression related to cell signaling, proliferation, and the production of cytoskeleton and ECM proteins.[108]

A specific role for biomaterial surface nanotopography has been demonstrated for growing nerve tissue. Control over the outgrowth of neurites from the cell bodies of neurons, both the sites at which they emerge from cell bodies and directions, was achieved *in vitro* using biomaterials with nanoscale groove dimensions.[104] The general conclusion of this study was that nanoscale substratum topography can be a potent morphogenetic factor for developing neurons and can assist in establishing neuronal polarity.

In addition to nanoscale texture, nanoscale roughness has been shown to affect cellular behavior. Using polymeric inverse replicas of native tissue, bladder smooth muscle cells[109] and cardiovascular endothelial cells[110] showed higher proliferation rates, more rapid spreading, and a more native-like appearance, respectively. Cells of osteogenic lineage also appear to be influenced by nanoscale topographies. In a series of studies using nanophase ceramics,[111–114] osteoblast (bone-forming) cells and osteoclast (bone-degrading) cells showed different behaviors dependent on nanophase ceramics. Proliferation, expression of differentiation markers, and calcium deposition were increased for osteoblasts. Similarly, the functions of osteoclast-like cells (including formation of resorption pits) were significantly enhanced.

2. *Protein and Peptide Nanostructures*

While synthetic surfaces lack specific signals cells can recognize, naturally derived materials may possess numerous signals involved in a wide variety of biological processes. In view of this characteristic, modulation of a biomaterial surface with nanoscale-sized, naturally occurring components of ECM proteins containing such signals has become a common modality. These nano approaches attempt to integrate living and nonliving systems. The effect of the immobilization of specific peptide sequences including RGD onto the surfaces of materials has been studied for biomaterials intended for implantation in both soft and hard tissues. Excellent reviews dealing with this topic[115,116] outline proven applications and promising potential for the future.

Biomaterial surfaces and matrices endowed with peptides for tissue engineering have been shown to enhance cellular behavior substantially (Table 3.3). Beneficial effects of such surfaces have been demonstrated for cell types with different functions and originating from both soft and hard tissues, i.e., connective tissues (fibroblasts), muscle tissues (myoblasts and smooth muscle cells), vascular tissues (endothelial cells), nerve tissues (neuronal cells), bone (osteoblasts), and cartilage (chondrocytes).

A pivotal factor determining the capacities of peptides to modulate cellular behavior is spatial distribution. In order to evoke functional intracellular signaling, receptors on a cell surface (e.g., integrins) must be clustered. This clustering can be achieved by increasing peptide density or flexibility using spacers. However, exaggerating the density of peptides on a biomaterial surface can also dramatically affect cell motility, which may be beneficial in immobilizing endothelial cells for vascular

Table 3.3 Application of Immobilized Peptides onto Biomaterials

Peptide	ECM Molecule Source	Application
RGD	Multiple ECM molecules, e.g., fibronectin, vitronectin, laminin, collagen, and thrombospondin	Enhance bone and cartilage tissue formation *in vitro* and *in vivo*; regulate neurite outgrowth *in vitro* and *in vivo*; promote myoblast adhesion, proliferation, and differentiation; enhance endothelial cell adhesion and proliferation
IKVAV YIGSR RNIAEIIKDI	Laminin	Regulate neurite outgrowth *in vitro* and *in vivo*
Recombinant fibronectin fragment (FNIII$_{7-10}$)	Fibronectin	Promote formation of focal contacts in preosteoblasts
Ac-GCRDGPQ-GIWGQDRCG	Common MMP substrates, e.g., collagen, fibronectin, and laminin	Encourage cell-mediated proteolytic degradation, remodeling, and bone regeneration *in vivo*

Source: Adapted from Boontheekul, T. and Mooney, D.J. *Curr Opin Biotechnol*, 14, 559–565, 2003.[137]

grafts, but is undesirable for the application of tissue ingrowth into tissue engineering matrices.

In vitro and *in vivo* studies investigating peptide-modulated biomaterials are numerous and focus primarily on RGD peptides. A comprehensive overview of some of these for use with both soft and hard tissue cells is presented below.

In vitro, RGD peptides significantly enhance attachment and spreading of cells, for example, endothelial cells.[117] Fibroblast proliferation rates were demonstrated to be significantly increased on RGD-modified polymeric material.[118] In addition to these cellular characteristics, cell migration and ECM deposition appear to be controllable using adequate density and distribution of RGD peptides.[119,120] In an attempt to enhance the attachment of endothelial cells to artificial blood vessels, polymeric materials were functionalized with a peptide domain selective for endothelial cells (and not for other cell types such as fibroblasts, vascular smooth muscle cells, and platelets).[121,122] Endothelial monolayers cultured on these functionalized polymeric materials proved to be nonthrombogenic, resulting in increased patency of such vascular grafts.

In vitro experiments have also clearly demonstrated the beneficial effects RGD immobilization can exert on cells of osteogenic lineage. Depending on the density and type of peptide immobilized on a material surface, both osteoblast-like cell adhesion and mineralization of synthesized ECM can be increased substantially.[72,123]

Tissue responses to peptide-treated polymeric material *in vivo* were assessed after intraperitoneal or subcutaneous implantation for 12 weeks.[124] Although blood sample analysis revealed no adverse responses, histological evaluation after 12 weeks demonstrated the presence of thicker fibrous capsules around RGD-treated implants compared to controls. In contrast, RGD-coated porous poly(methyl-methacrylate) implants in a rabbit model demonstrated enhanced and accelerated cancellous bone ingrowth compared to noncoated controls.[125] Moreover, apposition of newly formed bone directly toward the implant surface was observed, whereas noncoated control implants were surrounded by fibrous tissue layers that prevented direct bonding of bone to the implant surface.

3. Calcium Phosphate Nanostructures

A number of studies have described the effectiveness of calcium phosphate-coated implants and materials for bones. It is evident that calcium phosphate coatings improve the biological performance of endosseous implants.[6,126,127] Further expansion of calcium phosphate deposition techniques to generate coatings that resemble the nano-sized dimensions of native bone tissue have led to only a limited number of scientific publications to date.

Hydroxyapatite (HA) with nanometer-scale crystal size was used to modify commercially available collagen sheets.[128] Using an organ culture technique in which bone fragments were used to provide osteogenic cells, the composite scaffolds were demonstrated to be suitable for culturing osteogenic cells. The cells migrated from the bone fragments into the porous composite scaffold and eventually acquired a three-dimensional polygonal appearance. The three-dimensional osteogenic cells (nanoHA–collagen complexes) are suggested as promising candidates for the engineering of bone tissue. In a subsequent study using nanoHA–collagen composites, composite implants without bone cells (or fragments) were implanted into rabbit femurs.[129] Upon implantation, both new bone formation and implant degradation reminiscent of the bone remodeling process were observed. However, the lack of organization of bone constituents in the composite (compared to natural bone) resulted in a decreased mechanical strength that only reached the lower limit of that of natural bone.

V. CONSIDERATIONS

Although nanotechnologically modulated biomaterials have potential effects on cellular behavior, several issues must be considered. These issues involve determination of what actually causes specific cellular behavior and whether man-made nanoscale cues will be as powerful as natural nanoscale entities.

A. Topographical versus Chemical Cues

Roughly, all earlier described methods can be divided into techniques that (1) create isotropic nanotechnological structures that do not differ chemically from the intrinsic substratum or (2) create anisotropic nanotechnological structures using patterns of molecules chemically different from those of the intrinsic substratum.

Although both types of nanofabrication allow production of nanoscale cues on a biomaterial surface, the question of what causes a potential distinct cell behavior is justified since the former technique uses topographical cues to control cell behavior and the latter also uses chemical cues. Although chemical means to analyze possible slight differences in chemical properties of nanostructures on isotropic modified biomaterial surfaces are not yet available, topographical cues in the nanoscale range appear to affect the behavior of cells directly. The evidence for this is that the reactions of cells to similar topographies on chemically different biomaterial surfaces are comparable.[130]

Although we noted that the adsorption of proteins to biomaterial surfaces is responsible for cell attachment and subsequent cell reaction to biomaterial surfaces, the protein adsorption characteristics of materials still determine the constitution of the layer of proteins. Thus, if reactions of cells to topographical cues are similar regardless of the biomaterial used, it seems plausible to conclude that the adsorbed layer of proteins whose constitution is governed by the type of biomaterial[131] has a minor effect on cell reaction compared to the effect of the biomaterial surface topography.

Further evidence for this hierarchy of topographical over chemical cues in control of cell reactions has emerged from a study in which topography and chemistry competed in controlling alignments of neurites.[132] The biomaterial surfaces used in this experiment contained grooves or protein (laminin) patterns or both. Patterns were made either parallel or orthogonal. The conclusion drawn after culturing nerve cells on these substrates was that such morphogenetic guidance cues preferentially act synergistically. However, at sites where the depth of the grooves exceeded 500 nm, the topographical cues appeared to be higher in hierarchy than the chemical ones. It is important to emphasize that reactions of cells to either isotropic or anisotropic cues are dependent on cell type. Moreover, the presence of chemical cues is inevitably accompanied by minor parallel topographical cues since patterns of chemicals have certain thicknesses to which the cells may also react.

An *in vivo* model in which titanium implants were used showed that topography indeed is a major factor in modulating cell responses.[133] Polished, grit-blasted, HA-coated, and titanium film-covered HA-coated titanium implants were used to determine the relative contributions of surface topography and chemistry to the osseointegration of hard tissue implants. Using thin section histological evaluation and subsequent scanning electron microscopy, the authors showed that, although osseointegration was significantly greater on HA-coated titanium implants, 80% of the maximal bone forming response was observed on HA-coated titanium implants covered by a thin titanium film. The conclusion was that topography is a suprahierarchical factor compared to chemistry in bone apposition.

B. Natural versus Synthetic Nanostructures

The creation of nanostructures on biomaterial surfaces aimed to enhance implant success rates remains a man-made modification method. Therefore, even nanotechnological approaches to generating biomimetic biomaterial surfaces result in surfaces considerably dissimilar from those found in nature. The main reason is the lack of bilateral dynamic interactions and responses between the biomaterial surface and its biological surroundings. For example, the natural interaction of cell surface receptors (e.g., integrins) with their respective ligands is a highly dynamic process in which receptors and ligands continuously associate and disassociate, resulting in signaling into the cell via consecutive events of intracellular domains and accessory complexes and cascades.[134,135]

The clustering of receptor–ligand complexes is important for several cellular processes including motility[136] and the spatial distribution of receptor–ligand complexes is an important issue. Since the immobilization of ECM components on biomaterial surfaces does not allow spatial changes in ligand distribution upon cell binding, receptor clustering at the surfaces of cells adhering to the biomaterial surface is restricted to sites at which ligand density is appropriate after immobilization. Studies by Massia and Hubbell[75] aimed at elucidating the minimal distribution of RGD peptides required for interactions with the $\alpha_v\beta_3$–integrin receptor demonstrated that for human foreskin fibroblasts, a 440-nm peptide-to-peptide (RGD) spacing is required for cellular spreading, whereas the formation of focal contacts and stress fiber organization require an approximately three-fold lower RGD spacing.

VI. CONCLUSIONS

The generation of implants that will succeed in their intended functions requires a multidisciplinary approach that involves comprehension of divergent processes. For that reason, collaboration of researchers is recommended to develop clinically safe and reliable implants. When bulk properties of biomaterials meet the criteria required for a specific intended function, surface properties become important for minimizing potential undesirable responses by the surrounding biological environment. In view of this, we hope that nanotechnology may provide a significant approach by which a biomaterial surface can be modulated to decrease common host-versus-biomaterial responses.

Nanotechnology can provide strategies that can help create features on biomaterial surfaces in a dimensional range that may be adequate for cells. In its natural habitat, a cell is surrounded by other cells and by ECM proteins that provide a diverse range of signals (via cell–cell or cell–ECM contact) influencing cellular behavior. The majority of these signals are transmitted via receptor–ligand interactions, and their dimensions lie within the nanometer range. Therefore, several approaches using modulations of biomaterial surfaces with nanoscale features have been undertaken to study their effects on the responses of tissues in the direct vicinity of the implant. A wide variety of approaches include nanoscale topographical and chemical alterations at the biomaterial surface. Combinations of approaches (e.g., using both nanotopography and peptide functionalization) could offer additional power over cellular behavior.

Although research in this area is still in its infancy, several published studies indicate the beneficial effects nanotechnologically modified surfaces can have for implantology.[138] Because many aspects of cellular responses to materials are still unknown, further expansion of our understanding of nanotechnology and biological responses to nanoscale features will eventually result in clinically applicable designs for biomaterial surfaces that will be able to adjust to the required functionality of an implant.

The techniques described in this chapter are limited to experimental and laboratory settings. The clinical use of nanotechnologically modulated implants and prostheses awaits unambiguous proof of beneficial effects for given applications. However, improvements in exploiting currently available and future techniques combined with a better understanding of the influence of nanoscale features on cells and tissues surrounding implants and a multidisciplinary approach in implantology will pave the road for the use of nanostructures in the design of implants and prostheses.

REFERENCES

1. Anderson, J.M. et al. Implants and devices, in Ratner, B.D., Hoffman, A.S., Schoen, F.J., and Lemons, J.E., Eds., *Biomaterials Science: Introduction to Materials in Medicine*. Academic Press, London, 1996.
2. Ratner, B.D. Replacing and renewing: synthetic materials, biomimetics, and tissue engineering in implant dentistry. *J Dent Educ* 65, 1340–1347, 2001.
3. Crubezy, E., Murail, P., Girard, L., and Bernadou, J.P. False teeth of the Roman world. *Nature* 391, 29, 1998.
4. Cooke, F.W. Bulk properties of materials, in Ratner, B.D., Hoffman, A.S., Schoen, F.J., and Lemons, J.E., Eds., *Biomaterials Science: Introduction to Materials in Medicine*. Academic Press, London, 1996.
5. Black, J. *Biological Performance of Materials: Fundamentals of Biocompatibility*. Marcel Dekker, New York, 1992.
6. Lacefield, W.R. Materials characteristics of uncoated/ceramic-coated implant materials. *Adv Dent Res* 13, 21–26, 1999.
7. Schakenraad, J.M. Cells: their surfaces and interactions with materials, in Ratner, B.D., Hoffman, A.S., Schoen, F.J., and Lemons, J.E., Eds., *Biomaterials Sciences: Introduction to Materials in Medicine*. Academic Press, London, 1996.
8. Cordero, J., Munuera, L., and Folgueira, M.D. The influence of the chemical composition and surface of the implant on infection. *Injury* 27, Suppl. 3, SC34–SC37, 1996.
9. Bauer, T.W. and Schils, J. The pathology of total joint arthroplasty. II. Mechanisms of implant failure. *Skeletal Radiol* 28, 483–497, 1999.
10. Allara, D.L. Critical issues in applications of self-assembled monolayers. *Biosensors Bioelectron* 10, 771–783, 1995.
11. Williams, D.F. and Williams, R.L. Degradative effects of the biological environment on metals and ceramics, in Ratner, B.D., Hoffman, A.S., Schoen, F.J., and Lemons, J.E., Eds., *Biomaterials Science: Introduction to Materials in Medicine*. Academic Press, London, 1996.
12. Liu, W., Cao, Y., and Longaker, M.T. Gene therapy of scarring: a lesson learned from fetal scarless wound healing. *Yonsei Med J* 42, 634–645, 2001.
13. Webb, J.C.J. and Tricker, J. Bone biology: a review of fracture healing. *Curr Orthop* 14, 457–463, 2000.
14. Glowacki, J. Angiogenesis in fracture repair. *Clin Orthop* 355 (Suppl) S82–S89, 1998.
15. Harding, K.G., Morris, H.L., and Patel, G.K. Science, medicine and the future: healing chronic wounds. *Br Med J* 324, 160–163, 2002.
16. Singer, A.J. and Clark, R.A. Cutaneous wound healing. *New Engl J Med* 341, 738–746, 1999.

17. Bello, Y.M. and Phillips, T.J. Recent advances in wound healing. *JAMA* 283, 716–718, 2000.

18. Clark, R.A. Fibrin and wound healing. *Ann NY Acad Sci* 936, 355–367, 2001.

19. Hart, J. Inflammation. 1: Its role in the healing of acute wounds. *J Wound Care* 11, 205–209, 2002.

20. Hart, J. Inflammation. 2: Its role in the healing of chronic wounds. *J Wound Care* 11, 245–249, 2002.

21. Hollinger, J. and Wong, M.E. The integrated processes of hard tissue regeneration with special emphasis on fracture healing. *Oral Surg Oral Med Oral Pathol Oral Radiol Endod* 82, 594–606, 1996.

22. Wong, M.E., Hollinger, J.O., and Pinero, G.J. Integrated processes responsible for soft tissue healing. *Oral Surg Oral Med Oral Pathol Oral Radiol Endod* 82, 475–492, 1996.

23. Werner, S. and Grose, R. Regulation of wound healing by growth factors and cytokines. *Physiol Rev* 83, 835–870, 2003.

24. Gerstenfeld, L.C., Cullinane, D.M., Barnes, G.L., Graves, D.T., and Einhorn, T.A. Fracture healing as a post-natal developmental process: molecular, spatial, and temporal aspects of its regulation. *J Cell Biochem* 88, 873–884, 2003.

25. Baj-Krzyworzeka, M. et al. Platelet-derived microparticles stimulate proliferation, survival, adhesion, and chemotaxis of hematopoietic cells. *Exp Hematol* 30, 450–459, 2002.

26. Mackay, C.R. Chemokines: immunology's high impact factors. *Nat Immunol* 2, 95–101, 2001.

27. Janeway, C.A., Travers, P., Walport, M., and Capra, J. *Immunobiology: the immune system in health and disease.* Elsevier Science Ltd. (Garland Publishing), 1999.

28. Einhorn, T.A. The cell and molecular biology of fracture healing. *Clin Orthop* S7–S21, 1998.

29. Mutsaers, S.E., Bishop, J.E., McGrouther, G., and Laurent, G.J. Mechanisms of tissue repair: from wound healing to fibrosis. *Int J Biochem Cell Biol* 29, 5–17, 1997.

30. Visse, R. and Nagase, H. Matrix metalloproteinases and tissue inhibitors of metalloproteinases: structure, function, and biochemistry. *Circ Res* 92, 827–839, 2003.

31. Chamay, A. and Tschantz, P. Mechanical influences in bone remodeling: experimental research on Wolff's law. *J Biomech* 5, 173–180, 1972.

32. Thomsen, P. and Gretzer, C. Macrophage interactions with modified material surfaces. *Curr Opin Sol State Mater Sci* 5, 163–176, 2001.

33. Vogler, E.A. Structure and reactivity of water at biomaterial surfaces. *Adv Colloid Interface Sci* 74, 69–117, 1998.

34. Hallab, N.J., Bundy, K.J., O'Connor, K., Moses, R.L., and Jacobs, J.J. Evaluation of metallic and polymeric biomaterial surface energy and surface roughness characteristics for directed cell adhesion. *Tissue Eng* 7, 55–71, 2001.

35. Walboomers, X.F., Croes, H.J., Ginsel, L.A., and Jansen, J.A. Contact guidance of rat fibroblasts on various implant materials. *J Biomed Mater Res* 47, 204–212, 1999.

36. Jansen, J.A., van der Waerden, J.P., and de Groot, K. Effect of surface treatments on attachment and growth of epithelial cells. *Biomaterials* 10, 604–608, 1989.

37. Horbett, T.A. and Schay, M.B. Correlations between mouse 3T3 cell spreading and serum fibronectin adsorption on glass and hydroxyethylmethacrylate–ethylmethacrylate copolymers. *J Biomed Mater Res* 22, 763–793, 1988.

38a. Kasemo, B. and Gold, J. Implant surfaces and interface processes. *Adv Dent Res* 13, 8–20, 1999.

38b. Hanarp, P., Sutherland, D., Gold, J., and Kasemo, B. Nanostructured model biomaterial surfaces prepared by colloidal lithography. *Nanostructured Mater.* 12(1–4),
429–432, 1999.

39. Horbett, T.A. The role of adsorbed adhesion proteins in cellular recognition of
biomaterials. *BMES Bull* 23, 5–9, 1999.

40. Horbett, T.A. and Klumb, L.A. Cell culturing: surface aspects and considerations, in
Brash, W.P., Ed., *Interfacial Phenomena and Bioproducts.* Marcel Dekker, New York,
1996.

41. Steele, J.G., Dalton, B.A., Johnson, G., and Underwood, P.A. Adsorption of fibronectin and vitronectin onto Primaria and tissue culture polystyrene and relationship to
the mechanism of initial attachment of human vein endothelial cells and BHK-21
fibroblasts. *Biomaterials* 16, 1057–1067, 1995.

42. Steele, J.G. et al. Roles of serum vitronectin and fibronectin in initial attachment of
human vein endothelial cells and dermal fibroblasts on oxygen- and nitrogen-con-
taining surfaces made by radiofrequency plasmas. *J Biomater Sci Polym Ed* 6,
511–532, 1994.

43. Ratner, B.D. Reducing capsular thickness and enhancing angiogenesis around implant
drug release systems. *J Control Release* 78, 211–218, 2002.

44. Seare, W.J., Jr. Alloplasts and biointegration. *J Endourol* 14, 9–17, 2000.

45. Ducheyne, P. et al. Effect of hydroxyapatite impregnation on skeletal bonding of
porous coated implants. *J Biomed Mater Res* 14, 225–237, 1980.

46. Geesink, R.G., de Groot, K., and Klein, C.P. Bonding of bone to apatite-coated
implants. *J Bone Joint Surg Br* 70, 17–22, 1988.

47. Anderson, J. M. Biological responses to materials. *Annu Rev Mater Res* 31, 81–110,
2001.

48. van Diest, P.J., Beekman, W.H., and Hage, J.J. Pathology of silicone leakage from
breast implants. *J Clin Pathol* 51, 493–497, 1998.

49. Anderson, J.M. Multinucleated giant cells. *Curr Opin Hematol* 7, 40–47, 2000.

50. Zhao, Q. et al. Foreign body giant cells and polyurethane biostability: *in vivo* correlation of cell adhesion and surface cracking. *J Biomed Mater Res* 25, 177–183, 1991.

51. Whitesides, G.M. and Love, J.C. Nanofabrication: the art of building small. *Sci Am*
39–47, September 2001.

52. Sorribas, H., Padeste, C., and Tiefenauer, L. Photolithographic generation of protein
micropatterns for neuron culture applications. *Biomaterials* 23, 893–900, 2002.

53. Kane, R.S., Takayama, S., Ostuni, E., Ingber, D.E., and Whitesides, G.M. Patterning
proteins and cells using soft lithography. *Biomaterials* 20, 2362–2376, 1999.

54. Wilkinson, J.M. Nanotechnology: applications in medicine. *Med Dev Tech* 29–31,
June 2003.

55. ten Wolde, A. *Nanotechnology: Toward a Molecular Construction Kit.* Netherlands
Study Centre for Technology Trends, The Hague, 1998.

56. Curtis, A. and Wilkinson, C. Topographical control of cells. *Biomaterials* 18,
1573–1583, 1997.

57. Ito, Y. Surface micropatterning to regulate cell functions. *Biomaterials* 20, 2333–2342,
1999.

58. Desai, T.A. Micro- and nano-scale structures for tissue engineering constructs. *Med
Eng Phys* 22, 595–606, 2000.

59. Walboomers, X.F. *Engineered Implant Surfaces: Modification of Cell and Tissue
Response by Microgrooves.* University of Nijmegen, Netherlands, 2000.

60. Prokop, A. Bioartificial organs in the 21st century: nanobiological devices. *Ann NY
Acad Sci* 944, 472–490, 2001.

61. Yoshinari, M., Matsuzaka, K., Inoue, T., Oda, Y., and Shimono, M. Effects of mul-tigrooved surfaces on fibroblast behavior. *J Biomed Mater Res* 65A, 359–368, 2003.

62. von Recum, A.F., Shannon, C.E., Cannon, E.C., Long, K.J., van Kooten, T.G., and Meyle, J. Surface roughness, porosity, and texture as modifiers of cellular adhesion. *Tissue Eng* 2, 241–253, 1996.

63. Zhang, S. Emerging biological materials through molecular self-assembly. *Biotechnol Adv* 20, 321–339, 2002.

64. Whitesides, G.M., Mathias, J.P., and Seto, C.T. Molecular self-assembly and nanochemistry: a chemical strategy for the synthesis of nanostructures. *Science* 254, 1312–1319, 1991.

65. Mrksich, M., Dike, L.E., Tien, J., Ingber, D.E., and Whitesides, G.M. Using micro-contact printing to pattern the attachment of mammalian cells to self-assembled monolayers of alkanethiolates on transparent films of gold and silver. *Exp Cell Res* 235, 305–313, 1997.

66. Pierschbacher, M.D. and Ruoslahti, E. Cell attachment activity of fibronectin can be duplicated by small synthetic fragments of the molecule. *Nature* 309, 30–33, 1984.

67. Hynes, R.O. Integrins: versatility, modulation, and signaling in cell adhesion. *Cell* 69, 11–25, 1992.

68. Danen, E.H. and Sonnenberg, A. Integrins in regulation of tissue development and function. *J Pathol* 200, 471–480, 2003.

69. Kim, B.S. and Mooney, D.J. Development of biocompatible synthetic extracellular matrices for tissue engineering. *Trends Biotechnol* 16, 224–230, 1998.

70. Sakakura, S., Saito, S., and Morikawa, H. Stimulation of DNA synthesis in tropho-blasts and human umbilical vein endothelial cells by hepatocyte growth factor bound to extracellular matrix. *Placenta* 20, 683–693, 1999.

71. Barber, T.A., Golledge, S.L., Castner, D.G., and Healy, K.E. Peptide-modified pAAm-co-EG/AAc IPNs grafted to bulk titanium modulate osteoblast behavior *in vitro*. *J Biomed Mater Res* 64, 38–47, 2003.

72. Rezania, A. and Healy, K.E. Biomimetic peptide surfaces that regulate adhesion, spreading, cytoskeletal organization, and mineralization of the matrix deposited by osteoblast-like cells. *Biotechnol Prog* 15, 19–32, 1999.

73. Hubbell, J.A. Bioactive biomaterials. *Curr Opin Biotechnol* 10, 123–129, 1999.

74. Hubbell, J.A. Biomaterials in tissue engineering. *Biotechnology* 13, 565–576, 1995.

75. Massia, S.P. and Hubbell, J.A. An RGD spacing of 440 nm is sufficient for integrin alpha V beta 3-mediated fibroblast spreading and 140 nm for focal contact and stress fiber formation. *J Cell Biol* 114, 1089–1100, 1991.

76. Healy, K.E., Rezania, A., and Stile, R.A. Designing biomaterials to direct biological responses. *Ann NY Acad Sci* 875, 24–35, 1999.

77. Hill, I.R., Garnett, M.C., Bignotti, F., and Davis, S.S. Determination of protection from serum nuclease activity by DNA–polyelectrolyte complexes using an electro-phoretic method. *Anal Biochem* 291, 62–68, 2001.

78. Reyes, C.D. and Garcia, A.J. Engineering integrin-specific surfaces with a triple-helical collagen mimetic peptide. *J Biomed Mater Res* 65A, 511–523, 2003.

79. Cutler, S.M. and Garcia, A.J. Engineering cell adhesive surfaces that direct integrin alpha 5–beta 1 binding using a recombinant fragment of fibronectin. *Biomaterials* 24, 1759–1770, 2003.

80. Garcia, A.J. and Keselowsky, B.G. Biomimetic surfaces for control of cell adhesion to facilitate bone formation. *Crit Rev Eukaryot Gene Expr* 12, 151–162, 2002.

81. Barrere, F. et al. *In vitro* and *in vivo* degradation of biomimetic octacalcium phosphate and carbonate apatite coatings on titanium implants. *J Biomed Mater Res* 64A, 378–387, 2003.

82. Jansen, J.A., Wolke, J.G., Swann, S., Van der Waerden, J.P., and de Groot, K. Application of magnetron sputtering for producing ceramic coatings on implant materials. *Clin Oral Implants Res* 4, 28–34, 1993.

83. Wolke, J.G., van Dijk, K., Schaeken, H.G., de Groot, K., and Jansen, J.A. Study of the surface characteristics of magnetron-sputter calcium phosphate coatings. *J Biomed Mater Res* 28, 1477–1484, 1994.

84. Rivero, D.P., Fox, J., Skipor, A.K., Urban, R.M., and Galante, J.O. Calcium phosphate-coated porous titanium implants for enhanced skeletal fixation. *J Biomed Mater Res* 22, 191–201, 1988.

85. Leeuwenburgh, S., Wolke, J., Schoonman, J., and Jansen, J. Electrostatic spray deposition ESD of calcium phosphate coatings. *J Biomed Mater Res* 66A, 330–334, 2003.

86. McMichael, A.J. Antigens and MHC systems, in McGee, J.O.D., Isaacson, P.G., and Wright, N.A., Eds. *Oxford Textbook of Pathology.* Oxford University Press, Oxford, 1992.

87. Yamamoto, S. et al. DNA from bacteria, but not from vertebrates, induces interferons, activates natural killer cells and inhibits tumor growth. *Microbiol Immunol* 36, 983–997 1992.

88. Krieg, A.M. Immune effects and mechanisms of action of CpG motifs. *Vaccine* 19, 618–622, 2000.

89. Krieg, A.M. et al. CpG motifs in bacterial DNA trigger direct B-cell activation. *Nature* 374, 546–549, 1995.

90. Werner, M.H., Gronenborn, A.M., and Clore, G.M. Intercalation, DNA kinking, and the control of transcription. *Science* 271, 778–784 1996.

91. Wilson, W.D. Reversible interactions of nucleic acids with small molecules, in Blackburn, G.M. and Gait, M.J., Eds. *Nucleic Acids in Chemistry and Biology.* Oxford University Press, Oxford, 1996.

92. Goldman, A. and Glumoff, T. Interaction of proteins with nucleic acids, in Blackburn, G.M. and Gait, M.J., Eds. *Nucleic Acids in Chemistry and Biology.* Oxford University Press, Oxford, 1996.

93. Kamei, S., Tomita, N., Tamai, S., Kato, K., and Ikada, Y. Histologic and mechanical evaluation for bone bonding of polymer surfaces grafted with a phosphate-containing polymer. *J Biomed Mater Res* 37, 384–393, 1997.

94. Tretinnikov, O.N., Kato, K., and Ikada, Y. *In vitro* hydroxyapatite deposition onto a film surface-grated with organophosphate polymer. *J Biomed Mater Res* 28, 1365–1373, 1994.

95. Inoue, Y. et al. Antibacterial characteristics of newly developed amphiphilic lipids and DNA-lipid complexes against bacteria. *J Biomed Mater Res* 65A, 203–208, 2003.

96. Tanaka, K. and Okahata, Y. A DNA-lipid complex in organic media and formation of an aligned cast film. *J Am Chem Soc* 118, 10679–10683, 1996.

97. Okahata, Y. and Tanaka, K. Oriented thin films of a DNA–lipid complex. *Thin Solid Films* 284–285, 6–8, 1996.

98. Decher, G. Fuzzy nanoassemblies: toward layered polymeric multicomposites. *Science* 277, 1232–1237, 1997.

99. Brunette, D.M. and Chehroudi, B. The effects of the surface topography of micromachined titanium substrata on cell behavior *in vitro* and *in vivo*. *J Biomech Eng* 121, 49–57, 1999.

100. Singhvi, R. et al. Engineering cell shape and function. *Science* 264, 696–698, 1994.

101. Walboomers, X.F. and Jansen, J.A. Cell and tissue behavior on micro-grooved surfaces. *Odontology* 89, 2–11, 2001.

102. Parker, J.A., Walboomers, X.F., Von den Hoff, J.W., Maltha, J.C., and Jansen, J.A. Soft-tissue response to silicone and poly-L-lactic acid implants with a periodic or random surface micropattern. *J Biomed Mater Res* 61, 91–98, 2002.

103. den Braber, E.T., de Ruijter, J.E., Ginsel, L.A., von Recum, A.F., and Jansen, J.A. Quantitative analysis of fibroblast morphology on microgrooved surfaces with various groove and ridge dimensions. *Biomaterials* 17, 2037–2044, 1996.

104. Rajnicek, A., Britland, S., and McCaig, C. Contact guidance of CNS neurites on grooved quartz: influence of groove dimensions, neuronal age and cell type. *J Cell Sci* 110, Pt. 23, 2905–2913, 1997.

105. Clark, P., Connolly, P., Curtis, A.S., Dow, J.A., and Wilkinson, C.D. Cell guidance by ultrafine topography *in vitro*. *J Cell Sci* 99, Pt. 1, 73–77, 1991.

106. Wojciak-Stothard, B., Curtis, A., Monaghan, W., MacDonald, K., and Wilkinson, C. Guidance and activation of murine macrophages by nanometric scale topography. *Exp Cell Res* 223, 426–435, 1996.

107. Dalby, M.J., Riehle, M.O., Johnstone, H.J., Affrossman, S., and Curtis, A.S. Polymer-demixed nanotopography: control of fibroblast spreading and proliferation. *Tissue Eng* 8, 1099–1108, 2002.

108. Dalby, M.J. et al. Increasing fibroblast response to materials using nanotopography: morphological and genetic measurements of cell response to 13-nm-high polymer demixed islands. *Exp Cell Res* 276, 1–9, 2002.

109. Thapa, A., Miller, D.C., Webster, T.J., and Haberstroh, K.M. Nano-structured polymers enhance bladder smooth muscle cell function. *Biomaterials* 24, 2915–2926, 2003.

110. Goodman, S.L., Sims, P.A., and Albrecht, R.M. Three-dimensional extracellular matrix textured biomaterials. *Biomaterials* 17, 2087–2095, 1996.

111. Webster, T.J., Ergun, C., Doremus, R.H., Siegel, R.W., and Bizios, R. Enhanced osteoclast-like cell functions on nanophase ceramics. *Biomaterials* 22, 1327–1333, 2001.

112. Webster, T.J., Ergun, C., Doremus, R.H., Siegel, R.W., and Bizios, R. Enhanced functions of osteoblasts on nanophase ceramics. *Biomaterials* 21, 1803–1810, 2000.

113. Webster, T.J., Ergun, C., Doremus, R.H., Siegel, R.W., and Bizios, R. Specific proteins mediate enhanced osteoblast adhesion on nanophase ceramics. *J Biomed Mater Res* 51, 475–483, 2000.

114. Webster, T.J., Siegel, R.W., and Bizios, R. Osteoblast adhesion on nanophase ceramics. *Biomaterials* 20, 1221–1227, 1999.

115. Hersel, U., Dahmen, C., and Kessler, H. RGD-modified polymers: biomaterials for stimulated cell adhesion and beyond. *Biomaterials* 24, 4385–4415, 2003.

116. LeBaron, R.G. and Athanasiou, K.A. Extracellular matrix cell adhesion peptides: functional applications in orthopedic materials. *Tissue Eng* 6, 85–103, 2000.

117. Porte-Durrieu, M.C. et al. Development of RGD peptides grafted onto silica surfaces: XPS characterization and human endothelial cell interactions. *J Biomed Mater Res* 46, 368–375, 1999.

118. Davis, D.H., Giannoulis, C.S., Johnson, R.W., and Desai, T.A. Immobilization of RGD to <111> silicon surfaces for enhanced cell adhesion and proliferation. *Biomaterials* 23, 4019–4027, 2002.

119. Maheshwari, G., Brown, G., Lauffenburger, D.A., Wells, A., and Griffith, L.G. Cell adhesion and motility depend on nanoscale RGD clustering. *J Cell Sci* 113, Pt. 10, 1677–1686, 2000.

120. Mann, B.K. and West, J.L. Cell adhesion peptides alter smooth muscle cell adhesion, proliferation, migration, and matrix protein synthesis on modified surfaces and in polymer scaffolds. *J Biomed Mater Res* 60, 86–93, 2002.

121. Hubbell, J.A., Massia, S.P., Desai, N.P., and Drumheller, P.D. Endothelial cell-selective materials for tissue engineering in the vascular graft via a new receptor. *Biotechnology* 9, 568–572, 1991.

122. Massia, S.P. and Hubbell, J.A. Tissue engineering in the vascular graft. *Cytotechnology* 10, 189–204, 1992.

123. Rezania, A. and Healy, K.E. The effect of peptide surface density on mineralization of a matrix deposited by osteogenic cells. *J Biomed Mater Res* 52, 595–600, 2000.

124. Johnson, R. et al. Fibrous capsule formation in response to ultrahigh molecular weight polyethylene treated with peptides that influence adhesion. *Biomed Sci Instrum* 34, 47–52, 1997.

125. Kantlehner, M. et al. Surface coating with cyclic RGD peptides stimulates osteoblast adhesion and proliferation as well as bone formation. *Chem Biochem* 1, 107–114, 2000.

126. Dorozhkin, S.V. and Epple, M. Biological and medical significance of calcium phosphates. *Angew Chem Int Ed* 41, 3130–3146, 2002.

127. Kokubo, T., Kim, H.M., and Kawashita, M. Novel bioactive materials with different mechanical properties. *Biomaterials* 24, 2161–2175, 2003.

128. Du, C., Cui, F.Z., Zhu, X.D., and de Groot, K. Three-dimensional nano-HAp/collagen matrix loading with osteogenic cells in organ culture. *J Biomed Mater Res* 44, 407–415, 1999.

129. Du, C., Cui, F.Z., Feng, Q.L., Zhu, X.D., and de Groot, K. Tissue response to nano-hydroxyapatite/collagen composite implants in marrow cavity. *J Biomed Mater Res* 42, 540–548, 1998.

130. Curtis, A. and Wilkinson, C. Nantotechniques and approaches in biotechnology. *Trends Biotechnol* 19, 97–101, 2001.

131. Hlady, V. V. and Buijs, J. Protein adsorption on solid surfaces. *Curr Opin Biotechnol* 7, 72–77, 1996.

132. Britland, S. et al. Morphogenetic guidance cues can interact synergistically and hierarchically in steering nerve cell growth. *Exp Biol Online* 1, 2, 1996.

133. Hacking, S.A., Tanzer, M., Harvey, E.J., Krygier, J.J., and Bobyn, J.D. Relative contributions of chemistry and topography to the osseointegration of hydroxyapatite coatings. *Clin Orthop* 133, 24–38, 2002.

134. Zamir, E. and Geiger, B. Molecular complexity and dynamics of cell-matrix adhesions. *J Cell Sci* 114, 3583–3590, 2001.

135. Fernandez, C., Clark, K., Burrows, L., Schofield, N.R., and Humphries, M.J. Regulation of the extracellular ligand binding activity of integrins. *Front Biosci* 3, D684–D700, 1998.

136. Maheshwari, G., Brown, G., Lauffenburger, D.A., Wells, A., and Griffith, L.G. Cell adhesion and motility depend on nanoscale RGD clustering. *J Cell Sci* 113, Pt. 10, 1677–1686, 2000.

137. Boontheekul, T. and Mooney, D.J. Protein-based signaling systems in tissue engineering. *Curr Opin Biotechnol* 14, 559–565, 2003.

138. Castner, D.G. and Ratner, B.D. Biomedical surface science: foundations to frontiers. *Surf Sci* 500, 28–60, 2002.

CHAPTER 4

Diagnostics and High Throughput Screening

Aránzazu del Campo and Ian J. Bruce

CONTENTS

I. HIGH THROUGHPUT SCREENING AND NANOTECHNOLOGY TOOLS FOR BIOMEDICINE

A. Definition of High Throughput Screening

Molecular biology has become a common tool for research in medicine, particularly where investigations are undertaken to associate cellular dysfunction, disease, and therapeutic methods with specific perturbations at the molecular level. The vast range and complexity of biomolecules and the uniqueness of each individual's molecular (genetic or biochemical) profile make such investigations extremely complicated. Global screening of the entire molecular species of an organism or cell is an impossible task using traditional biological methods in which only analysis of selected components can be performed in parallel. In fact, considering disease and its underlying causes, selective screening may lead to false conclusions since changes in more than a single biomolecule are usually involved — simultaneously or consecutively — in the development of most illnesses.

This situation has encouraged researchers from many different disciplines to cooperate in the development of "global" analytical tools that allow the investigation of large sample collections in the context of biological targets in order to accurately identify active chemotypes. High throughput screening (HTPS) as we know it today at the start of the so-called "molecular age" of biomedicine is a consequence of this development.

HTPS approaches must fulfill certain criteria in order to be useful in a research or diagnostic laboratory. They must be able to perform large numbers of assays rapidly and simultaneously in a user-friendly manner and be small in format. They must be configured to provide robust and reproducible results that allow standardization and comparison of experiments performed in different laboratories. Because biological samples and reagents are usually small and costly to generate, HTPS methods should be capable of handling small volumes and detecting low concentrations of analytes in order to reduce cost. Finally, they should be capable of many reuses without significant reductions in accuracy or sensitivity.

Taking all these requirements into account, the goal for HTPS remains the fabrication of miniaturized laboratory reactors that can work in parallel and be compatible with high sensitivity detection systems to monitor their outputs. The emerging discipline of nanotechnology comes into play in this context by facilitating the creation of such systems through improved understanding and control of matter on a nanometer-length scale and its consequent exploitation.[1,2]

B. HTPS Architectures

Until now, microarray technology has represented the most widespread platform for HTPS in biomedical experimentation and diagnostics. Its impact is reflected in the increasing volume of scientific literature related to microarrays and their applications (Figure 4.1) and in the growth of the microarray market from $232 million in 1999 to $2.6 billion in 2004.[3]

Classic solid phase substrates such as microtiter plates, membrane filters, and microscopic slides used in biotesting inspired the development of microarrays. These media effectively represent flat substrates that can be modified so as to possess multiple (often hundreds or thousands) probe sites. Each site bears a ligand or probe whose molecular recognition of a complementary molecule can produce a signal that, when detected by an imaging technology, most often fluorescence, can indicate the interaction both quantitatively and qualitatively. These probe spots are micro- to nanometer-sized.

Microarrays can be classified on the basis of the materials arrayed upon them (Figure 4.2). They have been constructed using DNA and nucleic acids (natural and synthetic), proteins, antibodies, carbohydrates, tissues, and cells. Numerous

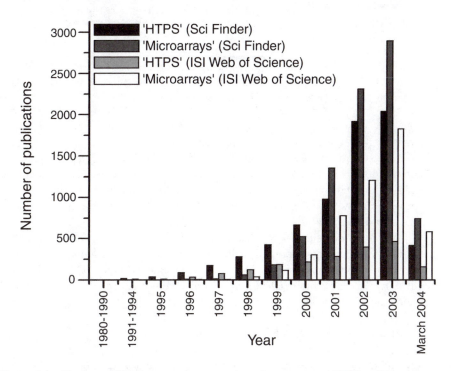

Figure 4.1 Number of literature reports concerning microarrays and HTPS published from 1980 to 2004. Data extracted from ISI Web of Science and SciFinder on-line search engines.

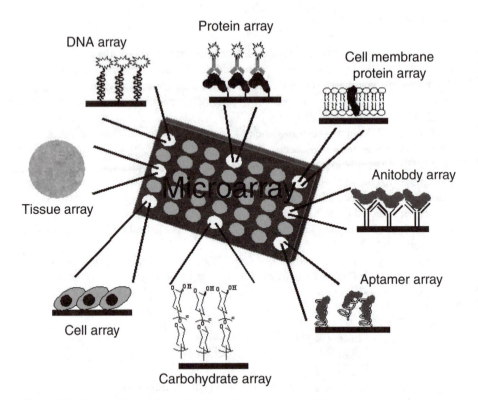

Figure 4.2   Types of flat surface microarrays by arrayed material.

examples of these arrays are commercially available, and descriptions of the most relevant formats will appear later in this chapter.

Newly emerging HTPS strategies are moving away from the classical formats. The new approaches may introduce significant benefits including diminished cost of fabrication and application and improvement in throughput. In this context, suspension arrays based on combinatorial libraries of encoded beads promise to enable ultra-HTPS analysis.[4]

Finally it is worth mentioning that microarrays would not be as effective as they are in HTPS without the help of microfluidics — a new term that defines any process or hardware involved in microvolume liquid management. In fact, HTPS and microfluidics overlap as commercial biosensors (biochips) often represent complicated networks of microsize channels, chambers, valves, and pumps.[5]

C. Nanotechnology and HTPS

Nanotechnology comes into play in the manufacture of microarrays and biochips because they benefit from the use of a great number of nanofabrication tools. Examples of nanotechnological contributions include (1) spatial positioning (microprinting and ink jetting) necessary for gridding arrayed materials in micro- to nanometer-scale spots; (2) patterning (photolithography) and microconstruction (micromachining, injection modeling, embossing) to fabricate channels and reservoirs for transporting fluids within a chip; and (3) surface molecular modification techniques (surface modification with self-assembled monolayers, surface spin coating with polymers or colloids) to control the properties of the array interface at the molecular level (adhesion, hydrophobicity, surface charge, friction). Detection systems must also be adjusted to take account of the nanometer dimensions and low concentrations of analytes involved in the processes.

For example, labeling strategies based on metal nanoparticles and quantum dots overcome the limitations of classical fluorescent chromophores and allow higher sensitivity and parallelism in detection.[6] New detection systems based on cantilevers, nanowires, and fiber optics are also expected to increase the precision and sensitivity of the detection process.[7] Finally, data handling, collection, and interpretation generated by HTPS including comparison and storage of databases will be absolutely necessary to take full advantage of HTPS. Therefore bioinformatics will be a key component of future HTPS developments.[8-10]

D. Principal Applications of HTPS in Biomedicine

HTPS is already changing the ways scientists and clinicians think of diseases and their treatments. Within the next two decades, HTPS is likely to replace present predominantly reactive approaches to disease diagnosis and monitoring with techniques that will be able to predict and prevent cellular dysfunction and illness.[11]

Global analysis of the genome, its transcription products (mRNA), and proteomes using DNA and protein arrays will help establish relationships between perturbations of common proteins or gene regulatory networks and disease states and development.[12] Rapid and inexpensive genome sequencing and polymorphism profiling will make it possible to make probabilistic statements about an individual's disease state or predisposition. Gene expression profile comparisons of individuals will also support these activities and make it possible for the first time to classify molecular variants of disease accurately.[13]

Efforts toward creating such predictive approaches are wasted if not accompanied by the development of suitable methods for preventing and combating disease. Therefore expression patterns obtained from mRNA "fingerprints" will also be used for identifying genes and pathways that are potential therapeutic targets. Protein microarrays will accelerate the discovery of peptide and protein therapeutics and the identification of candidate drug targets in pharmacognosy.[14] The use of arrays for HTPS in pharmacogenetics will open the door to individualized medicine that will take into account genetic markers for predispositioning to drug side effects and/or efficacy.[15]

Cell-based arrays will be used to test the efficacy and toxicity of drug candidates.[16] The power of cell-based assays combined with the other analytical arrays will enable the simplification, essentially the condensation, of serial drug discovery processes, thereby decreasing the time and cost of taking a "hit" compound to clinical trial. Such studies over time, when coupled with advances in our knowledge and understanding of proteins and nucleic acids, will provide extensive and diverse data structures and a new, more profound comprehension of cell biology that will be required for the truly predictive, preventive, and personalized medicine of the future.

II. FLAT SURFACE MICROARRAYS

A. DNA Microarrays

Glass slides, silicon wafers, and nylon membranes principally represent the architectures employed in flat surface DNA microarrays on which tens of thousands of different DNA sequences can be immobilized in ordered arrangements. Each arrayed spot is referred to as a feature, and features can be used to detect complementary DNA or mRNA sequences via hybridization interactions.[17] If a target sample is fluorescently labeled, automation of an array can allow the target sequence to be simultaneously detected and quantified upon its hybridization to the feature.

DNA microarrays can be classified on the basis of arrayed materials [complementary DNAs (cDNAs) or oligonucleotides (ONDs) or the fabrication method employed in their manufacture (spot arrays or *in situ* synthesized microarrays).[18–32] The arrayed material in cDNA arrays is usually a product of a polymerase chain reaction (PCR) generated from cDNA libraries or clone collections spotted onto glass slides or nylon membranes. Microarrays with short ONDs (15 to 25 mer) can be manufactured by *in situ* oligonucleotide synthesis onto silicon wafers or by spotting (50 to 120 mer) presynthesized oligonucleotides on glass slides.

The fabrication process involved in producing spotted microarrays includes three major steps: (1) chemical modification of the material arrayed in such a fashion that it can react with complementary functionalities present on the surface to form a stable bond (usually covalent); (2) coating of the support surface with adequate functional groups to allow specific covalent binding and prevent nonspecific adsorption of the arrayed material; and (3) use of a delivery system that brings small quantities (typically 50 to 100 nL) of the arrayed material to specific positions on the surface (printing or ink jetting).

Modification of synthetic ONDs for immobilization can now be performed during conventional automated synthesis. Moieties for use in linking can be added at the 3´ ends, at internal positions, or at the 5´ ends of such molecules, depending on the stage at which they are introduced in the synthetic pathway. Organosilanes carrying appropriate head groups capable of reacting directly with modified ONDs are commonly employed for surface activation of the supports involved. Amine, epoxy, carboxyl, or hydroxyl functionalities are the groups usually involved. Alternatively a cross-linking agent such as glutaraldehyde can be used.

In the case of cDNA arrays, slides coated with poly-lysine are most often used as substrates for grafting cDNAs. Spotted DNA adheres to such surfaces through electrostatic interactions. Hybridization with analyte DNA does not appear to disrupt this adhesion. Alternatively, covalent attachment of DNA to the surface may be achieved by photochemical cross-linking or by using primers modified with amine groups during the PCR reaction prior to spotting on an aldehyde- or carboxyl-modified surface.

Spotting cDNA or OND solutions onto a surface can involve bringing a "pin" or needle into contact with the surface (printing technology) or projecting a liquid droplet from a jet nozzle under pressure (ink jetting technology). A pin is essentially a highly miniaturized stainless steel fountain pen nib with a gap that draws up the spotting solution by capillary action. Pins can load up to 1 μL of solution and then deliver 50 to 100 nL spots upon contact with a surface. Usually multiple pins or jets integrated in a printhead are used to speed array production. Using these techniques, an array consisting of more than 30,000 spots with typical sizes between 75 and 500 μm in diameter can be fitted onto the surface of a conventional microscope slide.

A quality issue in spotting is the inhomogeneous distribution of the cDNAs or ONDs within each spot that can arise as a consequence of solvent evaporation — the so-called doughnut effect. Means to avoid this problem involve controlling environmental conditions during deposition (relative humidity) and changing the wettability properties of the drop by using a different solvent mixture for the solution.

Another issue is the spread of the spot after it is delivered on the surface. This can be avoided by modification of the surface with hydrophobic coating agents that can lead to an increase in surface tension and contact angle of the drop.

The fabrication of *in situ* synthesized OND microarrays involves light-directed, spatially addressable, parallel (combinatorial) chemical synthesis.[33] A surface is coated with linker molecules that bear photoremovable protecting groups or are covered by a photoresist layer. The pattern of irradiation (dictated by masks) deprotects (removes photoresist from) certain regions that are coupled to monomer units that are exposed. The process is repeated to build up different sequences at different sites. Such combinatorial synthesis enables 4^n different sequences of length n to be prepared in $4 \times n$ chemical steps. Using this method allows arrays of 1.6 cm^2 bearing up to 400,000 features to be prepared. In principle, the minimum feature size in these arrays is defined by the wave length of the irradiation source. However, in practice, current technologies have only produced features down to 25 μm.

In situ synthesized high-density OND arrays also differ from spotted arrays in the way that the target needs to be prepared for its quantitative determination. In both cases, genetic material from cells or tissue is extracted, amplified and fluorescently labeled. The 'tagged' nucleic acid sample is then added to the array and hybridizes to complementary ONDs therein. Using an appropriate fluorescence detection system, a quantitative two dimensional fluorescence image of hybridization intensity can rapidly be generated.

In situ synthesized OND chips allow direct determination of the number of hybridized sequences from the fluorescence intensity at each feature with a high degree of accuracy and reproducibility. Conversely, the process of gridding in spotted arrays is too inaccurate to allow comparisons of different arrays and it becomes necessary to mix and hybridize a reference nucleic acid to the same array, resulting in competitive binding of the target to the arrayed sequences. Target and reference nucleic acids are labeled with different fluorescent dyes and can be detected by scanning the array at two different λ_{max}^{em} wave lengths. Comparison of the fluorescence intensity of the target and reference nucleic acid at the same feature results in a quantitative measurement for each sequence represented on the array.[34]

In general, two key fabrication parameters limit the performance (in terms of throughput and sensitivity) of DNA microarrays: (1) the number of probe sites (features) per unit area that reflects its information density and (2) the number of probe molecules per unit area within an individual probe site that limits the number of target molecules that may bind to the array, and therefore the level of maximum sensitivity of the array.[33] In order to maximize the array throughput, the features and their spacing should be as small as possible while retaining full sensitivity and discrimination in terms of detection. Decreasing the size of the features has the additional advantage of reducing the amount of biological sample required in each analysis.

Several approaches have been adopted to increase the OND or cDNA density within a feature. One is based on the attachment to the surface of dendrimer molecules that act as multifunctional linkers and increase the density of the surface functional groups. An alternative method is to increase the surface area available by formation of porous 3D structures on the flat surface. Spin coating of gels onto the flat surfaces and deposition of porous thin films of nanometric colloidal silica particles are approaches used in this context. The increase in the surface area available in the porous structure permits the immobilization of a higher number of capture molecules per spot and leads to enhancement of hybridization signals (up to 20 times higher in the case of colloidal films).[35]

A number of issues must be considered in the selection of the most suitable format for a DNA array in the desired application. Spotted arrays have an advantage over *in situ* synthesized arrays in that they are easily customizable — they can be produced directly in the laboratory by individual investigators. Practically speaking, however, managing large clone libraries can be a daunting task for most laboratories, and making high quality spotted arrays can be difficult. OND microarrays offer advantages over cDNA microarrays: (1) greater specificity because immobilized sequences can be selected so as to represent only the specific part of the target sequence useful for hybrid capture and (2) the possibility to discern splice variants.[13] However, the complicated manufacturing processes involved in *in situ* OND arrays and the costs of producing large numbers of long ONDs required for spotted arrays make these arrays very expensive and limit their applicability for academic research groups.

An interesting example of the application of DNA arrays is in the analysis of protein–DNA interactions involving chromatin. These interactions are transient and therefore difficult to characterize in their physiological state.[36] The combination of chromatin immunoprecipitation assay (ChIP) with DNA array technology facilitates this type of analysis. In a ChIP–chip experiment, protein–DNA complexes are fixed in living cells by cross-linking with formaldehyde. Cell sonication shears the DNA into fragments and the target protein–DNA complexes are isolated by immunoprecipitation with a protein-specific antibody. The formaldehyde cross-links are then reversed and the DNA is purified, amplified by PCR, labeled with a fluorescent tag, and hybridized to a DNA microarray to identify the DNA regions bound to the protein.

One major challenge of DNA microarrays is their effective resolution of clinical questions.[37–40] In large part, this requires either the fabrication of custom arrays for the characterization of specific diseases or the fabrication of whole genome analysis arrays coupled to specific analytical methods that permit the analysis of the relevant genes for each disease. The need for low cost, technically simple arrays and easy-to-use analytical software for data collection and interpretation requires improvements to robotic and automation technologies for arraying features and processing arrays, new surface technology and chemistry for patterning and activation of glass slides and other substrates, and new labeling protocols and dyes.[13]

Another area for potential improvement lies in the related problem of the quantity of specimen needed for an analysis. Microarray experiments typically require between 10 and 40 µg of high quality nucleic acid to function. This corresponds roughly to a 100 mm^3 piece of tissue. The requirement can represent a large amount of material and ideally should be reduced. Finally, the degrees of sensitivity, specificity, accuracy, and reproducibility with microarray technology needed for accurate diagnosis in the medical arena are sometimes behind those achievable with PCR. Improved manufacturing efficiency, reduced experimental variation, and increased sensitivity will be essential for the future development of HTPS in this arena.

B. Protein Arrays

Protein array fabrication is much less developed as an art when compared to fabrication of DNA arrays.[41] The main reason is that DNA (and nucleic acids in general) can be represented as molecular species with relatively few, well defined chemical characteristics, but proteins are far more complex in structure and represent a wider range of chemical properties, and the ways in which proteins can interact with their target or complementary molecular species are many and varied and often depend closely on their tertiary and quaternary structures — and not simply their chemical compositions. Binding interactions can involve weak bonding forces, electrostatic interactions, H bonding, etc. In the case of nucleic acids, the primary consideration in this context is H bonding of pairing, complementary bases, representing a much simpler and more easily controlled situation.

Proteins therefore cannot be expected to behave as predictably as nucleic acids when immobilized on solid surfaces. In fact, certain inherent and useful properties may even be inactivated or absent when proteins are present as immobilized forms. In any case, the immobilization chemistry involved may lead to the inactivation of the very properties desirable in the application of the array. Chemical denaturation, physical conformation changes, and immobilization of the part of the molecule (epitope) required for the array application may all lead to dysfunctionality.

Another potential problem related to the construction and application of protein arrays is the ability of proteins to possess isoforms: the same protein can be present in different post-translationally modified variants and different forms of the same protein can arise as a consequence of gene splicing. It can therefore be extremely difficult to define a protein microarray and produce it in a form that will work effectively and reproducibly.

Finally, a further limiting factor is that no analog to PCR exists in the area of proteins. Thus, we cannot amplify the amount of protein for immobilization or detection in any convenient way. However, the potential for protein microarrays after these challenges are overcome is great. In fact, protein microarrays may be able to address applications currently impossible with nucleic acid basis approaches. One particular example is in epidemiological screening for immunity to infectious diseases. Different protein microarray platforms have now been developed and can be classified on the basis of the arrayed materials and their specific applications[42] as follows:

Interaction arrays — Low-density arrays of purified, native proteins used in the quantitative analysis of interactions with other proteins, nucleic acids, or small molecules.

Functional arrays — Arrays composed of purified, native proteins used in the prediction of protein function.

Affinity capture (protein expression) arrays — Arrays of affinity reagents capable of recognizing specific proteins and determining their presence and quantity in mixtures (see next section).

In the fabrication of protein arrays, material to be immobilized is usually robotically spotted onto the surfaces in question by microdispensing, liquid handling robots in a technique analogous to that for DNA arrays.[42,43] Alternatively, submicrometer feature arrays of proteins can be generated using "dip pen" nanolithography.[44,45] This method allows control of feature size <100 nm. This particular approach in HTPS formats still needs to be improved.

The supply of sufficient quantities of proteins is a major limiting factor in the construction and use of protein arrays.[46] The common methods for providing protein reagents for this purpose include cloning, expression, and parallel purification by affinity chromatography of the molecules involved. An alternative approach involves a combinatorial methodology with on-chip synthesis of features in a way similar to that for using *in situ* synthesized DNA arrays to generate high density peptide arrays.[47,48]

Most proteins are highly labile and susceptible to inactivation and/or conformational changes during modification and immobilization. As a consequence, considerable effort has been invested into optimizing the surface properties of supports in protein array technology in order to control the points of attachment and the densities of this class of ligand and to minimize nonspecific adsorption of proteins to the surface and denaturation of immobilized proteins.[49]

Proteins can be immobilized onto surfaces in nonoriented and oriented ways. The nonoriented technique usually involves random immobilization of the proteins onto a surface activated with functional groups capable of interaction with nonspecific functional groups on the protein, i.e., this could involve hydrophobic interactions between nitrocellulose-based polymers and proteins or covalent binding of amines, aldehydes, or epoxy groups on the substrate and free amine or carboxylic groups possessed by the protein. Such immobilization results in binding of a fraction of the protein in an orientation that impedes its interaction with the desired target.

Oriented immobilization potentially provides far better control over accessibility of target molecules to the active sites of the immobilized protein. This has clearly been demonstrated in the cases of (1) histidine (His)-tagged protein onto nitrilotriacetic acid-modified surfaces in the presence of Ni^{2+}, (2) biotinylated protein onto streptavidin monolayers, and (3) antibodies via oxidation of carbohydrate moieties on the Fc region and their conjugation via covalent bonding with surface hydrazide groups.

Another interesting class of protein array is composed of cell membrane proteins such as G protein-coupled receptors (GPCRs). Immobilization requires particular care so as to retain protein function. GPCRs in their native (functional) state are embedded in the phospholipid bilayer that forms the cell membrane, and this environment (together with the correct orientation of the protein within the membrane) is absolutely necessary for the GPCRs to retain their folded conformations and physiological roles as cell surface receptors responsible for transducing exogenous signals into intracellular responses. Therefore, GPCR array fabrication requires co-immobilization of the probe GPCR and its lipid membrane onto the array surface. Furthermore, the lipid membrane must be offset from the surface to avoid the physical contact that could otherwise induce misfolding or dysfunction of the extramembrane domains of the receptors. Covalent immobilization of the entire membrane is also undesirable because lateral mobility is an intrinsic and physiologically important property of native membranes.

Protein membrane microarrays have been fabricated in two different ways. The first approach uses the direct immobilization of membranes onto micropatterned substrates consisting of membrane-binding and nonmembrane-binding regions. The second approach uses microspotting technology by direct printing of solutions of membranes or membrane proteins onto membrane-binding surfaces.[50] Amine-modified surfaces are used for these purposes since they seem to provide the best combination of the requirements outlined above: preservation of protein conformation and orientation, lateral fluidity, and mechanical stability of the immobilized membrane.

The detection and quantification of the interaction of the arrayed immobilized proteins with their ligands is an important issue.[49] The preferred method for detecting binding events on a protein microarray relies on fluorescence. Fluorophores can be introduced into the assay via (1) direct fluorescent labelling of the sample to be tested or (2) sandwich immunoassays involving labelled antibodies. Direct labelling may seem the easiest and less expensive way to proceed, but an important consideration is that modification of proteins with fluorophores may alter their functions and/or structures. Also for multiplex binding assays, a cocktail of fluorescent labels may be required, thus making the assay in general and detection in particular more complex.

Two analytical techniques that do not require proteins to be labeled for their detection and that promise to be broadly useful for HTPS analysis of complex and undefined protein mixtures are matrix-assisted laser desorption/ionization time-of-flight mass spectrometry (MALDI-TOF-MS) and surface-plasmon resonance (SPR). MALDI-TOF-MS uses a laser pulse to desorb the immobilized proteins from the surface of the array, followed by identification of their molecular weights by mass spectroscopy. Detection by SPR spectroscopy relies upon the change of the mean refractive index of a surface that occurs when a protein is immobilized on it. This technique offers the advantage that it can be operated in solution — it does not require the substrate to be rinsed and dried before analysis and can therefore provide kinetic information on binding interactions. This is especially important for quantifying low affinity protein–protein interactions that are difficult to analyze using protocols where rinsing and drying are involved. However, these techniques have

Table 4.1 Comparison of Detection Methods for Analyzing Protein Arrays

Detection Method	Quantitative Analysis	Real-Time Analysis	Unlabeled Samples	Unbiased Assay	Availability
Fluorescence	Yes	No	No	No	High
Radiolabeling	Yes	No	No	No	High
MALDI-TOF	Semi	No	Yes	Yes	Medium
SPR	Yes	Yes	Yes	Partial	Limited

Note: Each method is qualitatively ranked for the following criteria: quantitative characterization of activity; real-time analysis of interactions; use of nonlabeled proteins and complex samples; identification of unanticipated activities; availability in research laboratories.

Source: From Lee, Y.S. and Mrksich, M. Trends Biotechnol, 20: S14–S18, 2002. With permission.

yet to prove their benefits for large scale protein profiling.[11] A comparison of the detection strategies is illustrated in Table 4.1.

C. Affinity Capture Arrays

Affinity capture arrays are composed of collections of immobilized affinity capture reagents. Affinity capture reagents are molecules that can interact specifically with a particular antigen or other specific molecular species by virtue of a recognition process. Antibodies are used most frequently as affinity capture reagents, although other molecular species such as antibody fragments, small globular proteins, small organic molecules, or aptamers (single-stranded ONDs with affinities for individual protein molecules) can be also used for this purpose. Reactive species and antigens may represent various molecular classes, ranging from biological (proteins, hormones) to nonbiological (certain drugs) molecules.

Antibodies (Abs) or immunoglobulins (Igs) are produced by an organism's immune system as part of the humoral immune response to a primary antigen stimulus. They are composed of four protein subunits: two identical polypeptide heavy chains (53 to 75 kD) and two identical light chains (~23 kD). These subunits are associated via disulfide bonds as well as by noncovalent interactions to form Y-shaped symmetrical dimers (Figure 4.3). The arms of this Y-shaped molecule contain the variable regions involved in antigen recognition and therefore form the active binding fragment (Fab). The stem of the Y (the crystallized fragment, Fc) contains the sites recognized by host defense mechanisms.

The particular molecular architecture and antigen recognition processes of Abs require that they are specifically oriented with respect to their surface attachments for optimal performance in assays. Their structure also lends itself for this purpose. For example, the heavy chains of Abs have N-linked oligosaccharides located in the Fc regions of the molecules. Oxidation of the hydroxyl groups of the sugar to aldehydes and covalent coupling to an amine- or hydrazide-modified array surface will yield an orientation of the immobilized Abs in which the nonactive Fc region is oriented toward the array surface and the Fab region outward toward the target protein. A similar effect can be obtained by array surface modification with molecules such as lectins that possess special affinities for carbohydrates.[51,52]

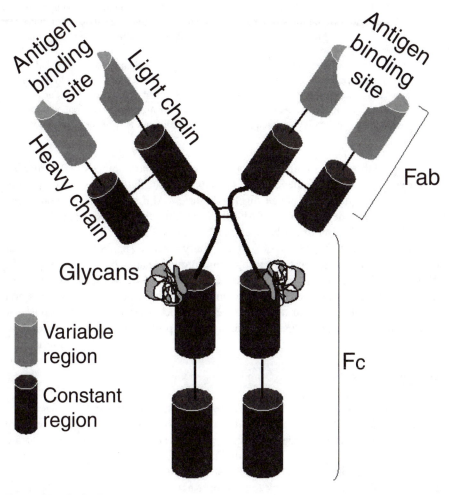

Figure 4.3 Antibody structure.

Oriented immobilization is also possible using sandwich structures. For example, the Fc portions of antibodies can be specifically recognized by proteins A and G. If the array surface is previously modified with one of these proteins, the subsequent immobilization of the Abs will proceed by the Fc region. Conversely, protein L interacts with the variable domain of Abs light chains and Abs immobilized in this case would preferentially expose the Fc region to the target.[52] Three main criteria are important in determining the success with which an affinity array can be used and applied:

1. The physical and physicochemical properties involved in the binding interaction (association/disassociation kinetics or k_m) stability of the complex formed under the array conditions and specificity of the recognition process)

2. The effect that immobilization can have on the affinity reagent (accessibility of those regions of the immobilized ligand to antigen, surface density of the affinity ligand, and inhibitory or negative effects that the immobilization chemistry may have on the ligand)

3. The total concentration of a particular antigen or other ligate in a mixture and its concentration relative to the others present (competitive effects of closely or distantly related ligate molecules for the affinity ligand; inhibitory effects at very low specific ligate concentrations when present in high concentrations of other molecules; substrative interactions between specific and nonspecific ligate molecules by sequestration)

For the fabrication of Ab arrays, a large quantity and number of Abs must be produced — routinely achieved through hybridoma technology or, if very large numbers are required, by phage display libraries. Although these methods have greatly improved the production of Abs, they are still quite expensive and this limits the use and application of antibody arrays.

As an alternative to the use of Abs as affinity reagents, nanofabricated artificial receptors capable of specifically recognizing protein shapes have been proposed.[2] Through molecular imprinting of polysaccharide-like films with the 3D shapes of protein molecules, engineered surfaces applicable to biochemical separations and assays have been generated. Such surfaces have been tested in the adsorption of proteins from single solutions or mixtures and the proteins adsorb preferentially to the positions on the surface where the complementary shape had been imprinted.

D. Carbohydrate Arrays

There are several reasons for studying glycans in biomedicine. (Glycans can be defined as carbohydrates that can be decomposed by hydrolysis into two or more molecules of monosaccharides.) Living organisms are composed of cells that are covered by diverse forms of glycans that help us to identify cell types and states. These glycans are involved in protecting cells against external physical stresses (e.g., freezing) and biochemical attack (e.g., proteases) and aid in cell–cell recognition, adhesion, and signaling — processes that are essential for normal tissue growth and repair as well as tumor cell transformation and metastasis.

Glycosylation is a form of post- or co-translational modification that occurs during eukaryotic protein synthesis. It is a key factor in determining a protein's function or dysfunction. Finally, cell surface glycans play a crucial role in bacterial and viral infections of host organisms. Microbes take advantage of these molecules to recognize and gain entry to host cells.

Despite their biological importance, the characterization of carbohydrate structures and the elucidation of their function have lagged behind characterization of proteins and nucleic acids. This is a consequence due in large part to the fact that polysaccharides in nature present a great structural diversity and this makes their study difficult. For example, polysaccharides may differ in terms of their monosaccharide residues, as well as in the types and molar ratios of the bonds linking them. These variations make it difficult to characterize them by sequencing, linkage determination, or 3D structural analysis. Sugar chain biosynthesis is complicated. Unlike the translation of mRNA into proteins, which is precisely mediated by the translational process, the biosynthesis of sugars and polysaccharides requires multiple enzymes and complex biosynthetic pathways. Also, polysaccharide functionality in living systems is strictly dependent upon their possession of specific and unique tertiary, and often quaternary structures. Their isolation and immobilization onto surfaces for microarray fabrication therefore require special care.

Various types of existing carbohydrate arrays can be differentiated on the basis of the molecular length (and consequent complexity) of the immobilized glycan. Arrays of monosaccharides, disaccharides, oligosaccharides, and carbohydrate-containing macromolecules (including polysaccharides and various glycoconjugate microarrays) have all been described.[53] The simplest formats composed of monosaccharides and disaccharides are suitable for preliminary screening and characterization of novel carbohydrate-binding proteins or carbohydrate-catalyzing enzymes and for identifying novel inhibitors of carbohydrate–protein interactions. However, certain proteins such as lectins and many Abs with anticarbohydrate reactivities can only recognize and bind to larger and more complex carbohydrate ligands or antigenic determinants. Monosaccharide and disaccharide sugar arrays are incapable of resolving investigations involving such molecular targets. In fact, oligosaccharide, polysaccharide, and glycoconjugate microarrays are used to perform this task.

The fabrication of glycan arrays can be performed by either *in situ* synthesis or by spotting carbohydrates onto activated supports. Various means, depending on the nature of the support and the type of glycan involved, can be used to attach carbohydrates to a support. Nitrocellulose-coated glass slides and nitrocellulose membranes have yielded particularly good results as supports in glycan microarray fabrication. The nitrocellulose polymer is a fully nitrated derivative of cellulose in which free hydroxyl groups are substituted by nitro groups, and the polymer is thus hydrophobic in character. It is still unclear why polysaccharides that are rich in hydroxyl groups and hydrophilic in nature should adsorb onto nitrocellulose supports. It has been suggested that the 3D microporous configuration of the nitrocellulose and the polymeric nature of the polysaccharides "fit" together to yield a particularly stable conformation of polysaccharides on a support. Nitrocellulose surfaces can be also used for the immobilization of glycoproteins. It is believed that immobilization occurs via interaction of hydrophobic regions of the protein with the membrane surface.

Covalent attachment of glycans to surfaces requires previous chemical modification of the carbohydrate molecules involved. Different ways to proceed in this context include: (1) attachment of biotinylated glycans to streptavidinated surfaces, (2) attachment of thiol-terminated polysaccharides to hydroxyl-terminated self-assembled monolayers, or (3) attachment of cyclopentadiene-terminated polysaccharides to quinone-terminated self-assembled monolayers by a Diels–Alder reaction.[53,54] In a manner analogous to the method for protein microarrays, the orientation of the immobilized glycan is important to the functionality of the array. For example, sugars must be displayed at the reducing end for successful protein recognition. However, practical access to sufficient carbohydrates of defined structure (either by isolation or synthesis) is a continuing problem.[55]

An issue still to be resolved for the future with respect to glycomics and analyzing interactions between carbohydrate binding proteins and oligosaccharides is how precisely the method can be used to determine "weak" affinities in such interactions. Most lectin–carbohydrate interactions are relatively weak and cannot be measured quantitatively with current technologies. From a biological perspective, this is probably important because cell–cell recognition events, supposedly mediated at least in part by lectins, are expected to be weak rather than strong. This could be particularly important in cancer studies.

SPR and microcantilever detection may provide the last hope where this type of analysis requiring extremely high sensitivity is required. However, the full potential for these techniques in HTPS has yet to be fully developed.

E. Cell Arrays

The types of arrays described above permit the assays of specific individual molecular interactions via HTPS but do not take into account the complex biology associated with whole living cells. Cell-based assays have been developed to permit such studies and allow automated monitoring of molecular processes within cells and cell function changes in a highly parallel manner.[56]

Different cell types in cell microarrays are spotted onto a support that has been modified to promote cellular adherence. Typical surface coatings to improve cell adherence are charged polymers such as poly-lysine or extracellular matrix components such as fibronectin or collagen. Coated substrates are commercially available or substrates can be prepared in-house at reasonable cost.

In order to increase data content and quality from HTPS with cell arrays, the arrays are designed to collect and analyze multiple data points from each feature in either multiparametric or multiplexed assays. Multiparametric cell assays, often called high content assays, permit analysis of multiple parameters from a single cell type. They are typically performed using automated platforms and high resolution microscopy to individually address the parameters to be measured. Multiplexed cell-based assays permit a single assay measurement for each cell type present at a probe site.[16] This type of cell assay has the advantage of a higher throughput than the multiparametric assay format, but possesses some potential limitations. First, the different cell types present must be able to grow or at least survive under a common set of conditions. Second, since the different cell types share the same extracellular environment, the possibility of cross-talk between them exists, and therefore measurements from the array could be compromised. Finally, the assay development required (technology and method) to multiplex a cell-based assay is unique. The signal to be analyzed from each individual cell type must be optimized so that under the same conditions it is possible to detect and quantify all the outputs simultaneously.

The most important consideration in the fabrication of cell arrays is the selection of the type of cell to be arrayed. In principle, primary cells (taken directly from a living organism) or transformed cell lines (cultures of a particular type of cell that is transformed so that it can grow and reproduce perpetually) can be selected. Primary cells of human origin are arguably the most physiologically relevant model systems for assays in the biomedical arena and human primary cell types are widely available commercially. However, in general, primary cells cannot be obtained on a scale necessary for HTPS and therefore transformed cell lines of human origin are the most commonly used cell-based HTPS platforms.

Cell lines can be also engineered to express or over-express a cDNA or protein of interest[57] and they can be used in the fabrication and production of so-called transfected cell microarrays. The fabrication of these microarrays is different from the description above and involves the printing of nanoliter quantities of cDNA-containing plasmids onto the surfaces of glass slides using a robotic microarrayer device. The printed arrays are then briefly exposed to a lipid transfection reagent, resulting in the formation of lipid–DNA complexes on the surfaces of the slides. Cells in medium are added on top of the arrayed cDNA, take up the plasmids, and become transfected. The arrays have important applications in drug discovery as a method of screening of gene products involved in biological processes of pharmaceutical interest and as *in situ* protein microarrays to aid in developing and assessing pharmaceutical compounds.

F. Tissue Microarrays

Large-scale human tissue analysis is crucial in many fields of medical research and diagnostics. This is particularly true for cancer research in which many different mechanisms can be involved in tumor development, as a result of which large numbers of tumors must be analyzed in studies to obtain a full representation of all genetic subtypes of a tumor type of interest. Previous methods for tissue analyses have been based either on homogenized tissue samples — a method that does not necessarily allow the specification of results to individual cell types — or the analysis of conventional tissue sections, which is a slow and tissue-consuming effort. Tissue microarrays that involve small sections of tissue samples arrayed onto glass slides significantly facilitate and accelerate this type of analysis.

The fabrication of tissue microarrays involves several steps (Figure 4.4).[58] First, core needle biopsies (typically 0.6 mm in diameter, 0.282 mm^2 surface area) are taken from a tissue donor block (paraffin-embedded tissue block or frozen tissue sample) and subsequently re-embedded into pre-made holes of an empty "recipient" paraffin block at a spacing between 0.2 and 0.8 mm (see figure). Regular microtomes are then used to cut sections from the recipient block and the sections then are transferred to a glass slide with the aid of an adhesive film.

A typical tissue array will possess about 600 samples per standard glass microscope slide, but new needles are under development that may allow as many as 2000 or more features per slide.[59] The final quality of the array is highly dependent on the dexterity of the individual constructing it, and it is particularly difficult to reproducibly generate standardized results for quantitative comparisons between tissues of the same array and even more difficult when considering comparison of different arrays even when constructed of the same materials. Controls from tissue samples or cell lines are usually placed on each array for comparative purposes and are necessary for the calibration of the array readers.

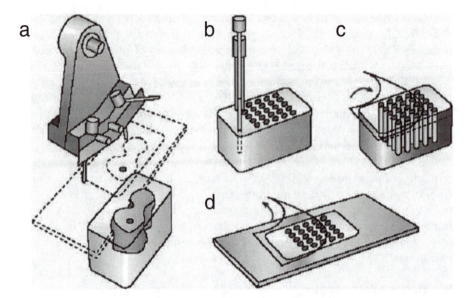

Figure 4.4   Tissue microarray fabrication. (a) Cylindrical tissue cores (usually 0.6 mm in diameter) are removed from a conventional (donor) paraffin block using a tissue microarrayer. (b) They are inserted into premade holes present in an empty (recipient) paraffin block. (c) Regular microtomes are used to cut tissue microarray sections. (d) The use of an adhesive-coated slide system facilitates the transfer of tissue microarray sections onto the slide and minimizes tissue loss, thereby increasing the number of sections that can be taken from each TMA block. (Photo couresy of Sauter, G., Simon, R., and Hillan, K. *Nat Rev Drug Discov* 2: 962–972, 2003.)

Tissue microarrays allow parallel detection of DNA or mRNA species by fluorescence *in situ* hybridization (FISH) and protein targets by immunohistochemistry (IHC).[60] However, automation of the tissue microarray reading process is currently a major factor limiting use. The reason is that any analysis must be performed in a truly representative area of the feature site. For example, if a microarray composed of tumor tissue is to be analyzed, the detection method must distinguish between measurements performed on malignant cells and those performed on nonmalignant tissue components (i.e., stroma, inflammation, or non-neoplastic epithelium) that may obscure the outcome of analysis.

Some methods appear to overcome this problem: (1) quantitative fluorescence image analysis (QIFA) that makes use of different fluorescence tags to differentiate cell types and define subcellular compartments and (2) simultaneous double direct immunofluorescence detection that makes use of one test and one reference antigen to normalize for the cellular content of detectable protein in each probe site. Although these methods improve the sensitivity of the assays, they also involve the development and evaluation of complex staining protocols — a time-consuming and expensive process. For these reasons, advances in nanoparticle staining and label-free detection systems (see next section) may move research in this area forward and aid in developing more sensitive detection systems capable of producing results with greater levels of reproducibility.

III. NONPOSITIONAL HTPS PLATFORMS

All the array systems discussed previously can be defined as positional. A feature of an array is defined in a 2D context (its x and y coordinates) with respect to a fixed or defined point on a slide determined by a reader. The detection of a signal at a particular x–y coordinate indicates that an event has occurred at that feature and from the intensity of the signal generated we can gain a quantitative idea of the amount of interaction that occurred. These types of arrays have limitations, including the difficulty with which they can be automated and fabricated, the volumes of samples required to permit them to function, their discriminatory abilities, and the complexities of the detection systems involved.

For these reasons, new approaches to fabricating and applying arrays are still being developed, some of which are nonpositional and do not rely on the spatial location of the feature to yield useful data. Among these alternative nonpositional approaches are the automated ligand identification system (ALIS), bead-based fiberoptic array, and suspension array.

A. Automated Ligand Identification System

ALIS is a nonpositional HTPS approach that permits the analysis of interactions of small molecules (that could be drug candidates) with particular target proteins on the basis of molecular weight measurement. The method starts with a library of hundreds to thousands of small organic compounds (potential drug candidates) in solution that is incubated with a target protein also in solution. After incubation, the solution is passed through a microscale size exclusion column that separates the protein and its bound ligands from the remaining library of molecules that have not interacted with the target.

The protein–ligand complex solution is then treated so as to dissociate the complex and the resulting solution is passed through a micro-reverse phase liquid chromatography column for concentration before it is fed into a mass spectrometer for structural identification of the ligands present. Since each ligand has a characteristic molecular mass, the analysis of the mass spectra of the mixture can reveal the identities of the ligands that interacted with the target. The drug candidates can be identified as those whose molecular weights match the peaks visible in the mass spectra. This platform can screen up to 300,000 compounds per day with minimal protein consumption and has been widely exploited in pharmacognosy.

B. Fiberoptic Arrays

Fiberoptic arrays are composed of bundles of thousands of fused optical fibers, each of them individually addressable and modified with a different molecular species that carries a specific fluorescent code permitting its specific detection.[31,61–63] Before describing these arrays, it is important to briefly review the basic principles of optical fibers.

An optical fiber (3 to 10 μm diameter) consists of a glass or plastic core surrounded by a cladding material. The fiber core can be selectively etched on one

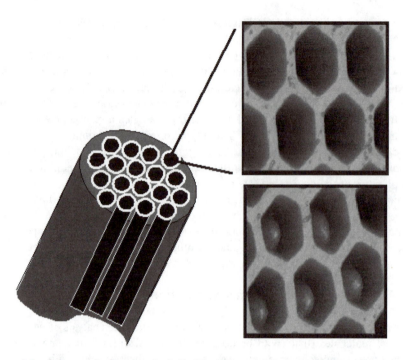

Figure 4.5   Schema of a fiber bundle (left). Atomic force micrographs of etched fiber bundles (top). Each well is 3 microns in diameter. The wells can be filled with complementary sized microspheres derivatized with different sensing chemistries (bottom). (Figures courtesy of Epstein, J.R. and Walt, D.R. *Chem Soc Rev* 32: 203–214, 2003.)

of its ends to form a sort of microwell capable of hosting molecular species, colloids, or even cells if modified with adequate surface chemistries (Figure 4.5). If the attached species are fluorescently labeled, the optical fiber can be also used as a fluorescence-based sensing tool when light at an appropriate excitation wave length is delivered through the fiber and the fluorescent indicator molecules fluoresce. The light emitted can be captured by the same fiber and transmitted back to a detector.

By fusing thousands of individual optical fibers into a densely packed bundle, an array of optical fibers can be constructed. This format has already been applied in the construction of DNA arrays in which a library of microspheres (encoding system) individually tagged with fluorophores, each carrying a specific OND at its surface, has been immobilized onto the core ends of the fibers.

This immobilization process at the core ends occurs randomly and positional registration of each sphere is necessary prior to the use of the array. Beaded optical fiber arrays differ markedly from the previously described positional arrays in that the position of each probe in the array is not registered by deliberate positioning during array fabrication, but is spectrally registered subsequent to its random distribution at the core tips. These arrays are used in a manner similar to that of positional arrays. The target molecules must be fluorescently labeled, and their fluorescence can be detected by the optical fibers in wells where hybridization has occurred.

Fiberoptic array platforms can also be used for fabrication of HTPS cell-based assays. Living cells are positioned in the etched wells of the core ends. The cells involved must be encoded with fluorophores to positionally register each specific cell type. By employing a range of fluorescent molecules or by varying the ratios of mixtures, multiple, different cell lines and strains can be addressed in parallel, permitting noninvasive and repetitive measurements of cell responses.

C. Suspension Arrays

Bead-based suspension arrays are becoming increasingly popular vehicles for screening and diagnostic applications. Addressable beads can be conjugated to ligands, oligonucleotides, or antibodies useful in a screening or diagnostic context. The beads are "bar coded" by incorporation of quantum dots, fluorophores, or even on the basis of size and physical structure so that they can be identified. The target molecule to be addressed can be also labeled and results are defined and confirmed in two ways: (1) in terms of the specific bead involved by confirmation of its identity and (2) confirmation that the interaction has occurred and its extent via the fluorescence signature of the target.[4,31]

Data collection and interpretation systems for handling results from these types of arrays can take various forms, depending on the bead bar coding method. In the case of fluorophores, flow cytometers are routinely involved. Alternatively, automated scanning confocal microscopy can be used. Regardless of encoding technique, these technologies produce arrays that are considerably more flexible and potentially more amenable to high throughput analysis than the positional technologies cited earlier. However, the powerful decoding methods capable of addressing each individual bead code necessary for HTPS are still currently in development.

IV. MICROFLUIDICS, MICROELECTROMECHANICAL SYSTEMS, AND MICRO TOTAL ANALYSIS SYSTEMS

Microfluidics is a developing technology involved in the transport and manipulation of minute amounts of fluids through microchannels that can be fabricated in a "chip" format (called micro- and nanoelectromechanical systems [MEMS and NEMS], respectively). With the help of microfluidics, the different steps involved in applying arrays to screenings or diagnostics can be integrated into small devices resembling miniaturized, automated laboratories (Figure 4.6).

This approach has been termed the micro total analysis system (μTAS) or lab-on-a-chip technology.[11,40,64] Such systems should contain elements for the pretreatment, separation, post-treatment, and detection of samples (Figure 4.7). The advantages of μTAS in diagnostics and HTPS include (1) improved performance, speed of analysis, and throughput; (2) reduced costs (minute sample volumes and reagent consumption); and (3) integration and multiplexing capabilities. Currently these micro and nano approaches still have certain analytical limitations, such as poor mixing efficiency, poor control of fluids in the microchannels, and low detection sensitivity. Considering the large impacts that fluctuations in small reaction volumes

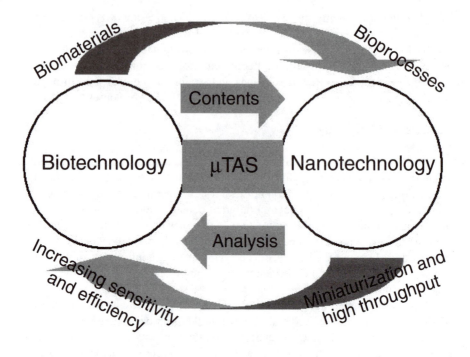

Figure 4.6 µTASs linking biology and nanotechnology. (From Lee, S.J. and Lee, S.T. *Appl Microbiol Biot* 64: 289–299, 2004. With permission.)

may have on analysis results, these features result in reduced reliability of tests conducted with these systems.

The fabrication of MEMS involves processes that are also common to the manufacture of microelectronic components, i.e., photolithography and surface micromachining to create structures with intricate details (vertical walls, chambers, freestanding beams or diaphragms, conduits, valves, etc.) and deposition of thin films to generate specialized surfaces for immobilization of biochemicals. Various µTAS[40,65–70] have been developed for the biomedical laboratory:

Microcapillary electrophoresis DNA chips for genomics — These arrays are constructed by using surface micromachining on glass, plastic, or silicon, to create a network of capillaries and reservoirs. Application of a voltage across such reservoirs causes fluid to flow along the microcapillaries. Analytes such as dissolved DNA fragments can be separated according to their electrophoretic mobility (a function of fragment length). Additional reservoirs connected by intersecting microcapillaries permit directional flow of the solution and hence processing of specific analytes to their respective "chemical stations."

PCR chips for genomics — These devices couple DNA analysis with *in situ* PCR for DNA amplification.[68]

Microcapillary electrophoresis chips for proteomics — These devices permit electrophoretic separation of proteins combined with mass spectroscopy detection through a microfabricated electrospray ionization source. As with protein microarray

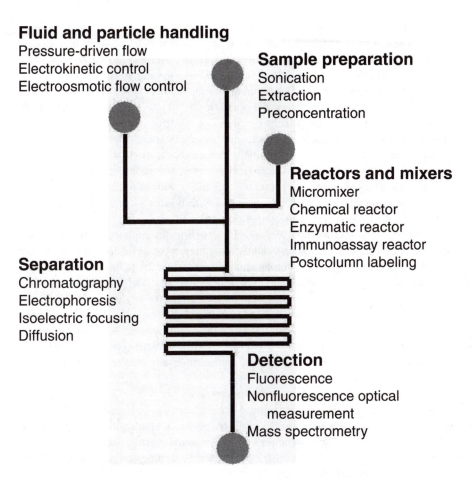

Fluid and particle handling
Pressure-driven flow
Electrokinetic control
Electroosmotic flow control

Sample preparation
Sonication
Extraction
Preconcentration

Reactors and mixers
Micromixer
Chemical reactor
Enzymatic reactor
Immunoassay reactor
Postcolumn labeling

Separation
Chromatography
Electrophoresis
Isoelectric focusing
Diffusion

Detection
Fluorescence
Nonfluorescence optical
   measurement
Mass spectrometry

Figure 4.7   Key technologies and components that must be incorporated in μTASs. (From Lee, S.J. and Lee, S.T. *Appl Microbiol Biot* 64: 289–299, 2004. With permission.)

systems, the technology for chip-based proteomic analysis is much less developed than that for genomics.

Microfluidic systems for analysis of mixtures of metabolites — These metabolites include glucose, uric acid, ascorbic acid, etc.

Cell-based chips for cellomics — These devices permit HTPS monitoring of physiological changes induced by exposure to environmental perturbations.

V. NEW TRENDS IN DETECTION SYSTEMS

A. New Labeling Systems: Nanoparticles and Quantum Dots

Recent nanotechnology advances allow access to a variety of nanostructured materials with unique optical properties. By manipulating structures at nanoscale dimensions, we can control and tailor the properties of materials at those dimensions, e.g., semiconductor nanocrystals and metal nanoshells, in a predictable manner to meet the needs of specific applications. In particular, nanotechnology may permit the development and application of optical imaging and biosensing by providing more robust contrast agents, fluorescent probes, and sensing substrates.

In addition, the size scale of such nanomaterials has benefits for many biomedical applications. The fact that many nanoparticles are similar in size (≤50 nm) to common biomolecules makes them potentially useful for intracellular tagging and makes them useful candidates for bioconjugate applications such as antibody targeting. In many cases, it is also possible to make modifications to nanostructures to better suit their integration with biological systems; for example, one may modify a surface in a way that enhances aqueous solubility, biocompatibility, or biorecognition. Nanostructures can also be embedded within other biocompatible materials to provide nanocomposites with unique properties.[2,6]

Why replace conventional molecular tags such as fluorophores with nanostructures? Current fluorescent markers can suffer from important inherent disadvantages including the requirement for color-matched lasers and the fading of fluorescence after even a single use. Also detection processes can lack discrimination when multiple dyes are employed in multiplex analyses due to the tendency of the different dyes to "bleed" together. Typically, nanostructured materials possess optical properties far superior to the molecular species they may replace — higher quantum efficiencies, greater scattering or absorbance cross-sections, optical activity over more biocompatible wave length regimes, and substantially greater chemical stability or stability against photobleaching.

Additionally, some nanostructures possess optical properties that are highly dependent on particle size or dimension. Such particles can be linked to biomolecules to form long-lived sensitive probes able to be used in identification processes. Successful examples of nanostructures that have been applied in detection processes in biotechnology and medicine are quantum dots, bioconjugated gold nanoparticles, and silver plasmon-resonant particles.

Quantum dots are highly light absorbing, luminiscent nanoparticles whose absorbance onset and emission maximum shift to higher energy with decreasing particle size due to quantum confinement effects.[6] Quantum dots are effectively nanocrystals typically in the size range of 2 to 8 nm in diameter. Unlike molecular fluorophores that typically have very narrow excitation spectra, semiconductor nanocrystals absorb light over a very broad spectral range. This makes it possible to optically excite a broad spectrum of quantum dot "colors" using a single excitation laser wave length that may enable one to simultaneously probe several markers in biosensing and assay applications. Moreover, the luminescence properties of quantum dots are also sensitive to their local environment and surface state. By using core-shell geometries where the nanocrystal is encapsulated in a shell of a wider band gap semiconductor, further improvements in the fluorescence quantum efficiencies (>50%) and photochemical stability of such materials have been achieved.

Applications of multicolor fluorescence imaging of arrays using quantum dots as a labeling system have been already reported,[71] and quantum dots can also be embedded within polymer-based nanoparticles or microparticles to bar code them for use in bead-based suspension arrays. A variety of colors and intensities of quantum dots can be employed for these purposes to generate what are effectively combinatorial libraries.

Gold colloidal nanoparticles have been also used for labeling target molecules as a consequence of their strong optical properties. They also possess the advantage of being highly versatile in terms of their bioconjugation; their surfaces can be modified easily with functional thiols to introduce active chemical groups capable of interacting with the biomolecules.[6]

A particular example of gold nanoparticle application has been in studies of biomolecular interactions (e.g., DNA hybridization, protein receptors) in which each species is conjugated to a gold particle and the complementary pairs can be discriminated on the basis of their different optical properties relative to each individual species. Because of the extremely strong optical absorption of gold colloids, this colorimetric method is sensitive enough to be able to detect down to 10 fmol of a labeled biomolecule. This method is approximately 50 times more sensitive than the sandwich hybridization detection methods used with molecular fluorophores.

Silver plasmon-resonant particles have been also used as labels in microarray-based DNA hybridization studies and sandwich immunoassays.[6] A particle consists of a gold nanoparticle core onto which a silver shell is grown and to which a biomolecule can be linked. Particles of this type in the size range of 40 to 100 nm have strong light scattering properties, allowing them to act as diffraction-limited point sources that can be observed using a standard dark-field microscope with white light illumination. When used as labels in immunoassays or hybridization assays, the results are determined by counting the number of particles bound to the substrate via microscopy. In DNA hybridization assays using such an approach, the detection sensitivity was approximately 60 times greater than that typically achieved using conventional fluorescent labels.

B. Label-Free Detection Systems

In recent years, significant effort has been dedicated to the development of label-free detection systems for use in HTPS and microarray systems. Detection approaches based on microcantilevers and nanowires have been described in the literature.

Microcantilevers measure the forces acting on a sharp tip as it approaches a specific target whose surface has been modified with a biomolecule (receptor).[72] Specifically, when the tip approaches the target, the nanomechanical forces act on it so as to cause the cantilever to bend. This bending can be detected by a laser that is capable of detecting deflections as small as a fraction of a nanometer. The deflection is proportional to the strength of the interaction, and thus also permits quantitative measurements to be made with these systems. This makes microcantilevers a potentially useful approach for HTPS.

Cantilevers are 0.01 the size of their macroscopic counterparts (quartz crystal microbalances) and can be mass-produced as miniaturized sensor arrays by current technologies. Silicon, silicon oxide, and nitride cantilevers are commercially available. They possess different shapes, dimensions, and force sensitivities capable of measuring in the 10^{-11} N range at the levels of single molecular interactions. Cantilever-based sensors are extremely versatile; they can be operated in air, vacuum, or liquid environments and transduce a number of different signals, e.g., magnetic, stress, electric, thermal, chemical, mass, and flow, into mechanical responses.

The receptor layer deposited on the cantilever surface directly affects its selectivity, reproducibility, and resolution. It is desirable to deposit a thin (to avoid changes in mechanical properties of the cantilever), uniform (to generate a uniform stress), and compact (to avoid interactions with the solid substrate beneath) layer of receptor molecules at the tip, and the surface coating should be stable and robust, with molecules covalently anchored to the surface while retaining enough freedom to interact with their specific ligand. This technology may be used to detect nucleic acid hybridization, antibody–antigen and receptor–ligand interactions, and enzymatic activity.[73–75]

Microcantilever arrays may be capable of assaying multiple proteins or nucleic acids in a single experiment or diagnostic test. Each cantilever surface would be prepared so that only one specific molecular species would be able to bind to it. As exciting as this prospect is, moving the technology from the initial proof-of-principle stage (where we are currently and where one cantilever is used at a time) to an array format (where several hundred cantilevers are represented on a single chip) is not trivial and requires much future work. Nonetheless, great enthusiasm surrounds such arrays, and results from early studies are promising.[76]

A further alternative to labels may be the use of semiconductor nanowires.[11] The idea is that a receptor molecule (antibody or single-stranded DNA) could be attached to a nanowire so that upon binding of the target species, measurable changes in the conductivity of the nanowire occurred. Such a detection device has the potential to be highly sensitive (in principle down to single molecule interactions) and one could imagine the construction of parallel arrays of nanowires where each one is functionalized with a different receptor. This system could also be used to make measurements in real time since it does not necessarily require labeling of the targets; rapid physiological processes (approximately 0.1 s in duration) could be measured. Integration of large numbers of these nanowires in a microfluidics device would be another potentially advantageous way of developing HTPS for the future.

VI. BIOINFORMATICS

The increasing amount and complexity of data arising from genetic, RNA expression, and proteomic screening led to a concomitant increase in potential usefulness. There is a clear requirement for computational analysis of such data and software to permit it. In fact, information technology has become an established component of basic industrial and academic research and product development to permit data mining of such information along with data collection and interpretation from HTPS systems.[8,9,11] DNA microarrays involve heavy reliance on computation[77] in:

Array design — Preselection of biological material to be printed in customs arrays (by using relational databases that include information from various sources, allowing efficient retrieval of biomolecules with the desired characteristics).

Image analysis — Quantification of the data displayed at each feature after scanning of the array (correct image capture, positional logging, precise detection of the features, feature centering, ability to detect at low quality features, background estimation and correction).

Storage and organization of experimental results — The potential to carry out thousands of experiments involving thousands of different genes needs effective database structures that can store the results of array experiments and facilitate data mining.

Comparison of screening profiles — Determination of groups possessing similar characteristics (e.g., clusters of genes); statistical analysis and interpretation of complex patterns of interacting groups to establish functional networks.[78]

VII. APPLICATIONS OF HTPS IN BIOMEDICINE

A. Genetic Diseases

Genetics and molecular medicine have benefited from rapid genotyping, mutational and polymorphism analysis, and DNA resequencing technologies.[79] Unlike conventional or classical approaches whose ability mainly resides in identifying individual genes whose expression is altered in the case of a particular disease,[17] microarray experiments are capable of identifying large numbers of genes whose expression is altered simultaneously or in a linked fashion as a consequence of disease. However, this often provides few clues as to which of the altered profiles are important in establishing a given phenotype (disease state). A given stimulus could potentially lead to changes in the expression levels of mRNA from hundreds of genes, particularly in mammalian systems.

In fact, the true power of DNA microarrays in elucidating genetic diseases has been illustrating global expression patterns rather than identifying single critical genes.[80–85] These expression profiles constitute a new tool for investigating patterns of diseases and identifying new disease genes for monogenetic disorders and complex traits, new functional and cellular relationships, and new pathways for the development of related drugs.[13] Literature concerning the application of microarrays in the determination and elucidation of genetic diseases is copious. Examples are in renal disease,[86–89] hepatic disease,[90] endocrinology and metabolism,[91–94] aging,[95] cardiovascular medicine,[96] oral, dental, and maxillofacial medicine,[97,98] otolaryngology and head and neck surgery,[99] muscle diseases,[100,101] rheumatic diseases,[85] and evolutionary theory.[102]

It is becoming increasingly clear that gene regulation depends not only on specific genetic composition, but also on epigenetic processes. An epigenetic process can be defined as one that relates to the conversion of genetic code information into a final product, i.e., transcription and translation, or any process involved in the interaction of genetic material. For example, in humans, the lack of promoter site methylation can lead to disease and abnormal DNA methylation is a hallmark of cancer cells. In fact, epigenetic silencing of tumor suppressor genes is thought to be a causal basis for a large number of sporadic human cancers. For these reasons, detection of methylation sites and the generation of methylation profiles with oligonucleotide microarrays are likely to be very important in future medicine.[103]

Expression patterns generated by DNA microarrays also provide us with important clues about the protein components of cells and tissues. Since mRNA and protein levels in eukaryotes are not necessarily directly correlated, it is necessary to co-analyze the protein complement of a cell at the same time as its mRNA to get a true indication as to the cellular changes associated with disease.[42] Therefore, protein and DNA arrays are complementary.[46]

The particular case of the central nervous systems of animals is an interesting one. It has been estimated that more than half of all genes are expressed at any one time and that many of them are rare low-abundance mRNAs that are more or less specific to this type of tissue.[104] The complexities of the nervous system at the level of the individual cells and their networks far surpass the complexities of other organ systems. Overlying this physical and physiological complexity is a diverse repertoire of functions that can change over time, for example, through aging.

The study of gene expression in the brain is of particular interest for a variety of reasons and it is particularly challenging. Microarrays are opening up this subject for the first time in a way that lets us begin to understand how and why tissues and systems exist and function. Reviews of the application of microarrays to neurobiology[105] cover genomic regulation of the brain,[106] the aging brain,[107] spinal cord injuries,[108] neurotoxicology,[109] schizophrenia and Alzheimer's disease,[110] and neurological disorders.[111]

As a particular example, we currently have little knowledge of which genes are involved in psychiatric disorders. In a pharmacognosy sense, notwithstanding the success achieved in recent years in developing new therapeutics based on receptor subtypes, a significant number of patients with these disorders remain resistant to treatment. Further, no systematic way permits determination of which of a variety of available treatments will be efficacious for a given patient. Limited progress has been made in identifying new and unique drug targets to treat particular illnesses. In the cases of psychiatric disorders such as schizophrenia and depression, it is clear that the conditions are caused by a set of abnormal genes and not by a single gene abnormality. Microarrays should allow us to identify the genes and allow us to successfully treat the conditions.[104,112]

B. Cancer

Cancer is a prime target to which array technologies can be addressed.[113–115] Several recent reviews describe the application of microarrays to the field of oncology, in particular, oral cancer,[116] ovarian cancer,[117–119] breast cancer,[113,120] hematologic malignancies,[121] and lymphomas.[122]

The various types of microarrays described earlier possess potential for application in cancer research and ultimately diagnosis and monitoring. For example, tissue microarrays can be applied for screening and comparison of the genetic and biochemical alterations occurring in different tumor tissues (multitumor tissue microarrays) and alterations occurring at different stages of tumor development (progression tissue microarrays). In addition, prognostic tissue microarrays containing samples of tumors from patients whose clinical follow-up data and endpoints are known can help identify novel prognostic parameters or link the chemotherapy responses of patients to alterations in their molecular profiles.[60,123] OND microarrays can be used in oncology for the detection of mutations or for the development of SNP fingerprints in populations of affected people that will help to better link heritable phenotypes to drug response (pharmacogenetics). cDNA microarrays can be used for screening for genomic imbalances (amplification of the oncogene or deletion of tumor suppressor gene).

In expression analysis, DNA microarrays permit the comparison of transcriptomes from normal and tumor tissues to clarify differences between normal and diseased phenotypes, provide comparisons of transcriptomes at various stages of cellular transformation for temporal assessment of tumor development, provide comparisons between transcriptomes from different samples of the same cancer type for classification of subtypes, and characterize transcriptome response to a variety of endogenous or exogenous interventions (pharmacogenomics and toxicogenomics).[124,125]

Additionally, proteomic studies with protein microarrays should help reveal the role of proteins in carcinogenesis and aid in the identification of protein fingerprints from which cancer biomarkers can be defined.[126,127] To date, microarray use in oncology has been restricted mainly to research; applications in routine clinical diagnosis and monitoring largely await the resolution of the following issues:

1. The amount of tumor tissue necessary for performing a DNA array experiment ($100 mm^3$) is usually too large to be obtained from formalin-fixed tissues. Tumor samples snap-frozen in liquid nitrogen immediately after surgical resection to prevent RNA degradation are ideally required, but are often difficult to obtain because of the constraints present in operating theaters. Moreover, biopsies intended for study tend to be small, increasingly so with the earlier detection of cancer and minimally invasive biopsy methods used currently.

2. Prospective identification, collection, and storage of high quality tissue are often lacking or poorly organized. This makes it especially difficult to make valid comparisons of data from different hospitals or research groups. A further complication is the fact that tissue quality can vary between locations (even between laboratories in the same institution) and the quality of nucleic acids, particularly RNA, extracted from tissues can vary dramatically. In addition, relevant clinical information regarding the tissues and specimens can be difficult to obtain in a retrospective fashion because of incomplete record keeping and patient confidentiality issues.

3. Sample selection problems. Tumors usually represent heterogeneous mixtures of different cell types, including malignant cells with varying degrees of differentiation, stromal elements, blood vessels, and inflammatory cells. Two tumors at similar clinical stages can vary markedly in grade and the relative proportions of different respective cell types. Tumors of different grades may differ in gene expression patterns, and different markers can be expressed either by malignant cells or by other cellular elements. This heterogeneity can complicate the interpretation of gene expression studies.

C. Genetic Epidemiology

Epidemiology involves the study of disease causes, distribution, and control in populations. It is largely an observational science whose objects of study are people who have and do not have diseases. Comparisons of these classes require the calculation of risks (i.e., probability of a disease based on exposure) and rates (i.e., frequency of disease per unit of population per unit of time). The science is based principally on statistical analyses and comparisons of populations of interest.

HTPS expression profiling with DNA arrays and the comparisons of expression patterns produced from different individuals will certainly help genetic epidemiologists by providing genetic markers in populations from which disease predisposition, diagnosis, and prognosis can be defined. Simultaneously, profiling and comparisons will permit for the first time the determination of whether broad genetic or environmental variations in populations account for the patterns of occurrence of disease, that is, the extent to which a disease is heritable; we may see the end of years of debate about which factors have the most influence on the occurrences of certain illnesses. This is especially important in cancer epidemiology since environmental variation is thought to dramatically influence the risk of cancer in certain cases.[103,128]

D. Tissue Typing

Tissue typing is particularly important for tissue matching during tissue and organ transplantation. Tissue rejection arising as a consequence of ill-matched tissues is a serious and potentially life threatening condition. Matching largely depends on the host's acceptance of the donor tissue in an antigenic context; in practice this has been achieved via human leukocyte antigen (HLA) typing.

However, the effectiveness of predicting graft rejection using this method is far from perfect. Increasing the resolution of tissue typing by the incorporation of single nucletide polymorphium (SNP) profiles of donors and recipients concerned about microarray analysis could potentially reduce graft rejection and as a result lessen the requirement for long-term, high-dose immunosuppression that carries significantly increased risks of morbidity and mortality for transplant recipients.[129] The development of SNP profiling via microarrays in clinical settings will require determination of appropriate SNP profiles that will provide the best matches for donor and recipient because some polymorphisms may be more immunogenic than others.

E. Infectious Diseases

The diagnosis and epidemiology associated with infectious diseases are additional areas where microarrays have been applied and offer benefits for the future.[3] Classical methods for microbial identification are often complex and time consuming and their replacement by quick and extremely sensitive multiplex screening assays is a major aim of clinical microbiology and public health laboratories. With the increasing number of microbial genomes that have been sequenced, our options for detailed analysis of infectious diseases and their causes have multiplied exponentially.[12] Many examples of microarray applications have been reported in the literature in the fields of clinical virology,[83,84,130,131] epidemiology,[131] infectious diseases,[132,133] vaccines,[134] parasitology,[135] malaria,[136] bacterial pathogenicity,[137] environmental microbiology,[3] and host–pathogen interactions.[138]

Carbohydrate microarrays may have a special role to play in the application of HTPS to infectious diseases since various microbes take advantage of cell membrane glycans to achieve infection of their host cells, and most bacterial toxins (cholera, diphtheria, tuberculosis) consist of sugar and carbohydrate-binding protein moieties (lectins).[53] Screening for protein–glycan interactions in HTPS format would be of great interest to pharmaceutical companies and national research institutes in terms of determining the structures and functions of such molecules and for drug development.

F. Therapeutics: Drug Discovery and Validation

Drug discovery is the process whereby compounds that exert activities against a specified target or function are identified, evaluated, and optimized for their performance in clinical and subclinical settings.[43,57,139–143] The process of drug discovery involves several steps, the first of which involves the identification and validation of a target, generally a gene product, whose function can be modulated by pharmacological products.

Examples could be compounds that inhibit the activities of gene products responsible for early brain swelling after a stroke or activation of the defective product of a mutant gene that causes a genetic disease. Typical goals include enzyme inhibitors, receptor agonists or antagonists, and transporter inhibitors or activators. Target identification and validation may involve gene and protein expression profiling using microarrays.[21]

In the second step, leading candidate compounds are identified by means of HTPS of diverse small molecule collections or structurally selected compounds with known or theoretically predicted activity against a target. "Hits" from this initial screening are then evaluated on the bases of many criteria, including but not limited to compound toxicity analysis, pharmacokinetics (compound distribution and metabolism in organs and bodily fluids, compound elimination, compound specificity, possible drug–drug interactions, mutagenic potential, and toxicity with long-term administration), and pharmacodynamics (efficacy *in vitro* and *in vivo*).[144,145] Secondary screening assays are used to confirm target specificity. The selected compounds are subjected to optimization by synthetic chemistry and more extensive preclinical evaluation in animal models.[146–150]

VIII. FUTURE OF NANOTECHNOLOGY AND HTPS

With the application of HTPS in clinical contexts, the world of diagnostics and therapeutics will truly enter the era of personalized medicine. Diagnosis, treatment, and management of patient conditions will be faster, more efficient, simpler, and more reliable than previously possible and therefore more beneficial to patients.

HTPS will permit tuning of therapeutics for optimized, individual patient treatment depending upon the patient's particular genotype.

Nanoscience and technology have central roles to play in this process. To achieve positive outcomes, bigger and more multidisciplinary research teams will be required to realize the anticipated revolution. In contrast to the time-honored models of academic collaboration among highly focused laboratories, nanoscience efforts will require that investigators learn each other's languages and form partnerships that integrate individual intellectual components into a cohesive team approach. Otherwise the very complexity of the new nanotechnologies will limit their applications in clinical and biomedical contexts.

REFERENCES

1. MC Roco. Nanotechnology: convergence with modern biology and medicine. *Curr Opin Biotechnol* 14: 337–346, 2003.
2. DF Emerich and CG Thanos. Nanotechnology and medicine. *Expert Opin Biol Th* 3: 655–663, 2003.
3. J Letowski, R Brousseau, and L Masson. DNA microarray applications in environmental microbiology. *Anal Lett* 36: 3165–3184, 2003.
4. KS Lam and M Renil. From combinatorial chemistry to chemical microarray. *Curr Opin Chem Biol* 6: 353–358, 2002.
5. J Khandurina and A Guttman. Microchip-based high throughput screening analysis of combinatorial libraries. *Curr Opin Chem Biol* 6: 359–366, 2002.
6. JL West and NJ Halas. Engineered nanomaterials for biophotonics applications: improving sensing, imaging, and therapeutics. *Annu Rev Biomed Eng* 5: 285–292, 2003.
7. T Vo-Dinh. Nanobiosensors: probing the sanctuary of individual living cells. *J Cell Biochem* 87: 154–161, 2002.
8. C Tilstone. Vital statistics. *Nature* 424: 610–612, 2003.
9. D Michalovich, J Overington, and R Fagan. Protein sequence analysis in silico: application of structure-based bioinformatics to genomic initiatives. *Curr Opin Pharmacol* 2: 574–580, 2002.
10. RJ Carroll. Variances are not always nuisance parameters. *Biometrics* 59: 211–220, 2003.
11. AD Weston and L Hood. Systems biology, proteomics, and the future of health care: toward predictive, preventative, and personalized medicine. *J Proteome Res* 3: 179–196, 2004.
12. G Walter, K Bussow, D Cahill, A Lueking, and H Lehrach. Protein arrays for gene expression and molecular interaction screening. *Curr Opin Microbiol* 3: 298–302, 2000.
13. RA Young. Biomedical discovery with DNA arrays. *Cell* 102: 9–15, 2000.
14. ZP Weng and C DeLisi. Protein therapeutics: promises and challenges for the 21st century. *Trends Biotechnol* 20: 29–35, 2002.
15. ME Chicurel and DD Dalma-Weiszhausz. Microarrays in pharmacogenomics: advances and future promise. *Pharmacogenomics* 3: 589–601, 2002.
16. OE Beske and S Goldbard. High-throughput cell analysis using multiplexed array technologies. *Drug Discov Today* 7: S131–S135, 2002.

17. A Schulze and J Downward. Navigating gene expression using microarrays: a technology review. *Nat Cell Biol* 3: E190–E195, 2001.
18. M Gabig and G Wegrzyn. An introduction to DNA chips: principles, technology, applications and analysis. *Acta Biochim Pol* 48: 615–622, 2001.
19. M Arcellana-Panlilio and SM Robbins. Cutting-edge technology. I. Global gene expression profiling using DNA microarrays. *Am J Physiol Gastr* 282: G397–G402, 2002.
20. MJ Heller. DNA microarray technology: devices, systems, and applications. *Annu Rev Biomed Eng* 4: 129–153, 2002.
21. CM Roth and ML Yarmush. Nucleic acid biotechnology. *Annu Rev Biomed Eng* 1: 265–297, 1999.
22. WM Freeman, DJ Robertson, and KE Vrana. Fundamentals of DNA hybridization arrays for gene expression analysis. *Biotechniques* 29: 1042–1045, 2000.
23. JD Pollock. Gene expression profiling: methodological challenges, results, and prospects for addiction research. *Chem Phys Lipids* 121: 241–256, 2002.
24. F Gros. From the messenger RNA saga to the transcriptome era. *Cr Biol* 326: 893–900, 2003.
25. H Zhu and M Snyder. "Omic" approaches for unraveling signaling networks. *Curr Opin Cell Biol* 14: 173–179, 2002.
26. DD Shoemaker and PS Linsley. Recent developments in DNA microarrays. *Curr Opin Microbiol* 5: 334–337, 2002.
27. HP Saluz, J Iqbal, GV Limmon, A Ruryk, and ZH Wu. Fundamentals of DNA chip/array technology for comparative gene expression analysis. *Curr Sci India* 83: 829–833, 2002.
28. L Smith and A Greenfield. DNA microarrays and development. *Hum Mol Genet* 12: R1–R8, 2003.
29. CM Ding and CR Cantor. Quantitative analysis of nucleic acids: the last few years of progress. *J Biochem Mol Biol* 37: 1–10, 2004.
30. D Vetter. Chemical microarrays, fragment diversity, label-free imaging by plasmon resonance: a chemical genomics approach. *J Cell Biochem* 87: 79–84, 2002.
31. BJ Battersby and M Trau. Novel miniaturized systems in high-throughput screening. *Trends Biotechnol* 20: 167–173, 2002.
32. A Bauer, B Beckmann, C Busold, O Brandt, W Kusnezow, J Pullat, V Aign, K Fellenberg, R Fleischer, A Jacob, M Frohme, and JD Hoheisel. Use of complex DNA and antibody microarrays as tools in functional analyses. *Comp Funct Genom* 4: 520–524, 2003.
33. MC Pirrung. How to make a DNA chip. *Angew Chem Int Edit* 41: 1277, 2002.
34. DV Nguyen, AB Arpat, NY Wang, and RJ Carroll. DNA microarray experiments: biological and technological aspects. *Biometrics* 58: 701–717, 2002.
35. M Glazer, J Fidanza, G McGall, and C Frank. Colloidal silica films for high-capacity DNA probe arrays. *Chem Mater* 13: 4773–4782, 2001.
36. MJ Buck and JD Lieb. ChIP chip: considerations for the design, analysis, and application of genome-wide chromatin immunoprecipitation experiments. *Genomics* 83: 349–360, 2004.
37. L Joos, E Eryuksel, and MH Brutsche. Functional genomics and gene microarrays: use in research and clinical medicine. *Swiss Med Wkly* 133: 31–38, 2003.
38. FL Kiechle and XB Zhang. The postgenomic era: implications for the clinical laboratory. *Arch Pathol Lab Med* 126: 255–262, 2002.
39. FL Kiechle and CA Holland-Staley. Genomics, transcriptomics, proteomics, and numbers. *Arch Pathol Lab Med* 127: 1089–1097, 2003.

40. RC McGlennen. Miniaturization technologies for molecular diagnostics. *Clin Chem* 47: 393–402, 2001.

41. P Cutler. Protein arrays: current state of the art. *Proteomics* 3: 3–18, 2003.

42. MF Templin, D Stoll, JM Schwenk, O Potz, S Kramer, and TO Joos. Protein microarrays: promising tools for proteomic research. *Proteomics* 3: 2155–2166, 2003.

43. C Huels, S Muellner, HE Meyer, and DJ Cahill. The impact of protein biochips and microarrays on the drug development process. *Drug Discov Today* 7: S119–S124, 2002.

44. JM Nam, SW Han, KB Lee, XG Liu, MA Ratner, and CA Mirkin. Bioactive protein nanoarrays on nickel oxide surfaces formed by dip-pen nanolithography. *Angew Chem Int Edit* 43: 1246–1249, 2004.

45. KB Lee, SJ Park, CA Mirkin, JC Smith, and M Mrksich. Protein nanoarrays generated by dip-pen nanolithography. *Science* 295: 1702–1705, 2002.

46. G Walter, K Bussow, A Lueking, and J Glokler. High-throughput protein arrays: prospects for molecular diagnostics. *Trends Mol Med* 8: 250–253, 2002.

47. XL Gao, XC Zhou, and E Gulari. Light directed massively parallel on-chip synthesis of peptide arrays with t-Boc chemistry. *Proteomics* 3: 2135–2141, 2003.

48. R Frank. The SPOT synthesis technique: synthetic peptide arrays on membrane supports: principles and applications. *J Immunol Methods* 267: 13–26, 2002.

49. YS Lee and M Mrksich. Protein chips: from concept to practice. *Trends Biotechnol* 20: S14–S18, 2002.

50. Y Fang, J Lahiri, and L Picard. G protein-coupled receptor microarrays for drug discovery. *Drug Discov Today* 8: 755–761, 2003.

51. SP Lal, RI Christopherson, and CG dos Remedios. Antibody arrays: an embryonic but rapidly growing technology. *Drug Discov Today* 7: S143–S149, 2002.

52. G Elia, M Silacci, S Scheurer, J Scheuermann, and D Neri. Affinity-capture reagents for protein arrays. *Trends Biotechnol* 20: S19–S22, 2002.

53. DN Wang. Carbohydrate microarrays. *Proteomics* 3: 2167–2175, 2003.

54. KR Love and PH Seeberger. Carbohydrate arrays as tools for glycomics. *Angew Chem Int Edit* 41: 3583–3586, 2002.

55. J Hirabayashi. Oligosaccharide microarrays for glycomics. *Trends Biotechnol* 21: 141–143, 2003.

56. PA Johnston. Cellular platforms for HTS: three case studies. *Drug Discov Today* 7: 353–363, 2002.

57. SN Bailey, RZ Wu, and DM Sabatini. Applications of transfected cell microarrays in high-throughput drug discovery. *Drug Discov Today* 7: S113–S118, 2002.

58. G Sauter, R Simon, and K Hillan. Tissue microarrays in drug discovery. *Nat Rev Drug Discov* 2: 962–972, 2003.

59. DL Rimm, RL Camp, LA Charette, J Costa, DA Olsen, and M Reiss. Tissue microarray: new technology for amplification of tissue resources. *Cancer J* 7: 24–31, 2001.

60. H Moch, J Kononen, OP Kallioniemi, and G Sauter. Tissue microarrays: what will they bring to molecular and anatomic pathology? *Adv Anat Pathol* 8: 14–20, 2001.

61. JR Epstein, APK Leung, KH Lee, and DR Walt. High-density, microsphere-based fiber optic DNA microarrays. *Biosens Bioelectron* 18: 541–546, 2003.

62. JR Epstein and DR Walt. Fluorescence-based fibre optic arrays: a universal platform for sensing. *Chem Soc Rev* 32: 203–214, 2003.

63. DR Walt. Molecular biology: bead-based fiber-optic arrays. *Science* 287: 451–452, 2000.

64. AD Sheehan, J Quinn, S Daly, P Dillon, and R O'Kennedy. The development of novel miniaturized immuno-sensing devices: review of a small technology with a large future. *Anal Lett* 36: 511–537, 2003.

65. SJ Lee and SY Lee. Micro total analysis system (mu-TAS) in biotechnology. *Appl Microbiol Biot* 64: 289–299, 2004.

66. J Khandurina and A Guttman. Bioanalysis in microfluidic devices. *J Chromatogr A* 943: 159–183, 2002.

67. J Khandurina and A Guttman. Microscale separation and analysis. *Curr Opin Chem Biol* 7: 595–602, 2003.

68. LJ Kricka and P Wilding. Microchip PCR. *Anal Bioanal Chem* 377: 820–825, 2003.

69. P Gascoyne, J Satayavivad, and M Ruchirawat. Microfluidic approaches to malaria detection. *Acta Trop* 89: 357–369, 2004.

70. TH Park and ML Shuler. Integration of cell culture and microfabrication technology. *Biotechnol Progr* 19: 243–253, 2003.

71. JK Jaiswal, H Mattoussi, JM Mauro, and SM Simon. Long-term multiple color imaging of live cells using quantum dot bioconjugates. *Nat Biotechnol* 21: 47–51, 2003.

72. LA Bottomley. Scanning probe microscopy. *Anal Chem* 70: 425–475, 1998.

73. P Dutta, CA Tipple, NV Lavrik, PG Datskos, H Hofstetter, O Hofstetter, and MJ Sepaniak. Enantioselective sensors based on antibody-mediated nanomechanics. *Anal Chem* 75: 2342–2348, 2003.

74. JH Pei, F Tian, and T Thundat. Glucose biosensor based on the microcantilever. *Anal Chem* 76: 292–297, 2004.

75. M Su, SU Li, and VP Dravid. Microcantilever resonance-based DNA detection with nanoparticle probes. *Appl Phys Lett* 82: 3562–3564, 2003.

76. B Daviss. A springboard to easier bioassays. *Scientist* 18: 40–44, 2004.

77. J Tamames, D Clark, J Herrero, J Dopazo, C Blaschke, JM Fernandez, JC Oliveros, and A Valencia. Bioinformatics methods for the analysis of expression arrays: data clustering and information extraction. *J Biotechnol* 98: 269–283, 2002.

78. TG Dewey. From microarrays to networks: mining expression time series. *Drug Discov Today* 7: S170–S175, 2002.

79. M Carella, S Volinia, and P Gasparini. Nanotechnologies and microchips in genetic diseases. *J Nephrol* 16: 597–602, 2003.

80. PA Clarke, RT Poele, R Wooster, and P Workman. Gene expression microarray analysis in cancer biology, pharmacology, and drug development: progress and potential. *Biochem Pharmacol* 62: 1311–1336, 2001.

81. V Reinke and KP White. Developmental genomic approaches in model organisms. *Annu Rev Genom Hum G* 3: 153–178, 2002.

82. G Colebatch, B Trevaskis, and M Udvardi. Functional genomics: tools of the trade. *New Phytol* 153: 27–36, 2002.

83. RO Elferink. One step further toward real high-throughput functional genomics. *Trends Biotechnol* 21: 146–147, 2003.

84. F Michiels, H van Es, and P Tomme. One step further toward real high throughput functional genomics. *Trends Biotechnol* 21: 147–148, 2003.

85. S Steer and AJ MacGregor. Genetic epidemiology: disease susceptibility and severity. *Curr Opin Rheumatol* 15: 116–121, 2003.

86. K Susztak, K Sharma, M Schiffer, P McCue, E Ciccone, and EP Bottinger. Genomic strategies for diabetic nephropathy. *J Am Soc Nephrol* 14: S271–S278, 2003.

87. P Devarajan, J Mishra, S Supavekin, LT Patterson, and SS Potter. Gene expression in early ischemic renal injury: clues toward pathogenesis, biomarker discovery, and novel therapeutics. *Mol Genet Metab* 80: 365–376, 2003.

88. PS Hayden, A El-Meanawy, JR Schelling, and JR Sedor. DNA expression analysis: serial analysis of gene expression, microarrays and kidney disease. *Curr Opin Nephrol Hy* 12: 407–414, 2003.

89. EP Bottinger, WJ Ju, and J Zavadil. Applications for microarrays in renal biology and medicine. *Exp Nephrol* 10: 93–101, 2002.

90. NA Shackel, MD Gorrell, and GW McCaughan. Gene array analysis and the liver. *Hepatology* 36: 1313–1325, 2002.

91. F Eertmans, W Dhooge, S Stuyvaert, and F Comhaire. Endocrine disruptors: effects on male fertility and screening tools for their assessment. *Toxicol in Vitro* 17: 515–524, 2003.

92. KD Hirschi, JA Kreps, and KK Hirschi. Molecular approaches to studying nutrient metabolism and function: an array of possibilities. *J Nutr* 131: 1605s–1609s, 2001.

93. GP Page, JW Edwards, S Barnes, R Weindruch, and DB Allison. A design and statistical perspective on microarray gene expression studies in nutrition: the need for playful creativity and scientific hard-mindedness. *Nutrition* 19: 997–1000, 2003.

94. E Bernal-Mizrachi, C Cras-Meneur, M Ohsugi, and MA Permutt. Gene expression profiling in islet biology and diabetes research. *Diabetes Metab Res* 19: 32–42, 2003.

95. S Welle. Gene transcript profiling in aging research. *Exp Gerontol* 37: 583–590, 2002.

96. SA Cook and A Rosenzweig. DNA microarrays: implications for cardiovascular medicine. *Circ Res* 91: 559–564, 2002.

97. WP Kuo, ME Whipple, ST Sonis, L Ohno-Machado, and TK Jenssen. Gene expression profiling by DNA microarrays and its application to dental research. *Oral Oncol* 38: 650–656, 2002.

98. WP Kuo, ME Whipple, TK Jenssen, R Todd, JB Epstein, L Ohno-Machado, ST Sonis, and PJ Park. Microarrays and clinical dentistry. *J Am Dent Assoc* 134: 456–462, 2003.

99. ME Whipple and WP Kuo. DNA microarrays in otolaryngology: head and neck surgery. *Otolaryng Head Neck* 127: 196–204, 2002.

100. EP Hoffman, KJ Brown, and E Eccleston. New molecular research technologies in the study of muscle disease. *Curr Opin Rheumatol* 15: 698–707, 2003.

101. JN Haslett and LM Kunkel. Microarray analysis of normal and dystrophic skeletal muscle. *Int J Dev Neurosci* 20: 359–365, 2002.

102. AY Gracey and AR Cossins. Application of microarray technology in environmental and comparative physiology. *Annu Rev Physiol* 65: 231–259, 2003.

103. B van Steensel and S Henikoff. Epigenomic profiling using microarrays. *Biotechniques* 35: 346, 2003.

104. Z Luo and DH Geschwind. Microarray applications in neuroscience. *Neurobiol Dis* 8: 183–193, 2001.

105. TJ Sendera, D Dorris, R Ramakrishnan, A Nguyen, D Trakas, and A Mazumder. Expression profiling with oligonucleotide arrays: technologies and applications for neurobiology. *Neurochem Res* 27: 1005–1026, 2002.

106. SJ Watson, F Meng, RC Thompson, and H Akil. The "chip" as a specific genetic tool. *Biol Psychiatr* 48: 1147–1156, 2000.

107. TA Prolla. DNA microarray analysis of the aging brain. *Chem Senses* 27: 299–306, 2002.

108. FM Bareyre and ME Schwab. Inflammation, degeneration and regeneration in the injured spinal cord: insights from DNA microarrays. *Trends Neurosci* 26: 555–563, 2003.

109. KE Vrana, WM Freeman, and M Aschner. Use of microarray technologies in toxicology research. *Neurotoxicology* 24: 321–332, 2003.
110. ER Marcotte, LK Srivastava, and R Quirion. cDNA microarray and proteomic approaches in the study of brain diseases: focus on schizophrenia and Alzheimer's disease. *Pharmacol Therap* 100: 63–74, 2003.
111. JC Weeks. Thinking globally, acting locally: steroid hormone regulation of the dendritic architecture, synaptic connectivity and death of an individual neuron. *Progr Neurobiol* 70: 421–442, 2003.
112. WE Bunney, BG Bunney, MP Vawter, H Tomita, J Li, SJ Evans, PV Choudary, RM Myers, EG Jones, SJ Watson, and H Akil. Microarray technology: a review of new strategies to discover candidate vulnerability genes in psychiatric disorders. *Am J Psychiatr* 160: 657–666, 2003.
113. DG Albertson. Profiling breast cancer by array CGH. *Breast Cancer Res Tr* 78: 289–298, 2003.
114. S Mohr, GD Leikauf, G Keith, and BH Rihn. Microarrays as cancer keys: an array of possibilities. *J Clin Oncol* 20: 3165–3175, 2002.
115. S Ramaswamy and TR Golub. DNA microarrays in clinical oncology. *J Clin Oncol* 20: 1932–1941, 2002.
116. JK Nagpal and BR Das. Oral cancer: reviewing the present understanding of its molecular mechanism and exploring the future directions for its effective management. *Oral Oncol* 39: 213–221, 2003.
117. CA Bandera, B Ye, and SC Mok. New technologies for the identification of markers for early detection of ovarian cancer. *Curr Opin Obstet Gyn* 15: 51–55, 2003.
118. I Haviv and IG Campbell. DNA microarrays for assessing ovarian cancer gene expression. *Mol Cell Endocrinol* 191: 121–126, 2002.
119. W Wu, W Hu, and JJ Kavanagh. Proteomics in cancer research. *Int J Gynecol Cancer* 12: 409–423, 2002.
120. F Bertucci, P Viens, P Hingamp, V Nasser, R Houlgatte, and D Birnbaum. Breast cancer revisited using DNA array-based gene expression profiling. *Int J Cancer* 103: 565–571, 2003.
121. TR Golub. Genomic approaches to the pathogenesis of hematologic malignancy. *Curr Opin Hematol* 8: 252–261, 2001.
122. C Schwaenen, S Wessendorf, HA Kestler, H Dohner, P Lichter, and M Bentz. DNA microarray analysis in malignant lymphomas. *Ann Hematol* 82: 323–332, 2003.
123. M van de Rijn and CB Gilks. Applications of microarrays to histopathology. *Histopathology* 44: 97–108, 2004.
124. AE Frolov, AK Godwin, and OO Favorova. Differential gene expression analysis by DNA microarray technology and its application in molecular oncology. *Mol Biol* 37: 486–494, 2003.
125. JKC Chan. The new World Health Organization classification of lymphomas: the past, the present and the future (reprinted from *Int. Med* 86, 434–443, 2000). *Hematol Oncol* 19: 129–150, 2001.
126. JE Celis and P Gromov. Proteomics in translational cancer research: toward an integrated approach. *Cancer Cell* 3: 9–15, 2003.
127. BB Haab. Methods and applications of antibody microarrays in cancer research. *Proteomics* 3: 2116–2122, 2003.
128. DD Dalma-Weiszhausz, ME Chicurel, and TR Gingeras. Microarrays and genetic epidemiology: a multipurpose tool for a multifaceted field. *Genet Epidemiol* 23: 4–20, 2002.

129. A Boussioutas and I Haviv. Current and potential uses for DNA microarrays in transplantation medicine: lessons from other disciplines. *Tissue Antigens* 62: 93–103, 2003.

130. P Kellam. Post-genomic virology: the impact of bioinformatics, microarrays and proteomics on investigating host and pathogen interactions. *Rev Med Virol* 11: 313–329, 2001.

131. JP Clewley. A role for arrays in clinical virology: fact or fiction? *J Clin Virol* 29: 2–12, 2004.

132. PA Bryant, D Venter, R Robins-Browne, and N Curtis. Chips with everything: DNA microarrays in infectious diseases. *Lancet Infect Dis* 4: 100–111, 2004.

133. D Ivnitski, DJ O'Neil, A Gattuso, R Schlicht, M Calidonna, and R Fisher. Nucleic acid approaches for detection and identification of biological warfare and infectious disease agents. *Biotechniques* 35: 862–869, 2003.

134. FX Berthet, T Coche, and C Vinals. Applied genome research in the field of human vaccines. *J Biotechnol* 85: 213–226, 2001.

135. JC Boothroyd, I Blader, M Cleary, and U Singh. DNA microarrays in parasitology: strengths and limitations. *Trends Parasitol* 19: 470–476, 2003.

136. PK Rathod, K Ganesan, RE Hayward, Z Bozdech, and JL DeRisi. DNA microarrays for malaria. *Trends Parasitol* 18: 39–45, 2002.

137. GK Schoolnik. Functional and comparative genomics of pathogenic bacteria. *Curr Opin Microbiol* 5: 20–26, 2002.

138. M Kato-Maeda, Q Gao, and PM Small. Microarray analysis of pathogens and their interaction with hosts. *Cell Microbiol* 3: 713–719, 2001.

139. DN Howbrook, AM van der Valk, MC O'Shaughnessy, DK Sarker, SC Baker, and AW Lloyd. Developments in microarray technologies. *Drug Discov Today* 8: 642–651, 2003.

140. D Brunner, E Nestler, and E Leahy. In need of high throughput behavioral systems. *Drug Discov Today* 7: S107–S112, 2002.

141. TW Gant. Application of toxicogenomics in drug development. *Drug News Perspect* 16: 217–221, 2003.

142. AS Verkman. Drug discovery in academia. *Am J Physiol Cell Ph* 286: C465–C474, 2004.

143. EH Ohlstein, RR Ruffolo, and JD Elliott. Drug discovery in the next millennium. *Annu Rev Pharmacol* 40: 177–191, 2000.

144. KV Chin and ANT Kong. Application of DNA Microarrays in pharmacogenomics and toxicogenomics. *Pharmaceut Res* 19: 1773–1778, 2002.

145. G Orphanides. Toxicogenomics: challenges and opportunities. *Toxicol Lett* 140: 145–148, 2003.

146. H Loferer. Mining bacterial genomes for antimicrobial targets. *Mol Med Today* 6: 470–474, 2000.

147. M Basik, S Mousses, and J Trent. Integration of genomic technologies for accelerated cancer drug development. *Biotechniques* 35: 580, 2003.

148. KD Kumble. Protein microarrays: new tools for pharmaceutical development. *Anal Bioanal Chem* 377: 812–819, 2003.

149. TW Snell, SE Brogdon, and MB Morgan. Gene expression profiling in ecotoxicology. *Ecotoxicology* 12: 475–483, 2003.

150. JI Glass, AE Belanger, and GT Robertson. *Streptococcus pneumoniae* as a genomics platform for broad-spectrum antibiotic discovery. *Curr Opin Microbiol* 5: 338–342, 2002.

CHAPTER 5

Nano-Enabled Components and Systems for Biodefense

Calvin Shipbaugh, Philip Antón, Gabrielle Bloom,
Brian Jackson, and Richard Silberglitt

CONTENTS

I. INTRODUCTION

The classes of products that are applicable strictly for biodefense may also be useful against various molecular-based threats (e.g., chemical agents) as well as biological threats. This chapter will review many examples of nanotechnology that may lead to components and systems — including methods based on biological components — with practical applications for defense of human health, security against biological warfare or terrorism, agriculture, and the environment. Figure 5.1 illustrates that the response requires several steps beginning with the sensor and characterization sequence.

The molecular natures of chemical toxins and dangerous biological materials make nanotechnology an obvious choice for developing defenses to counter these hazards. Figure 5.2 characterizes the potential benefits of nanotechnology. It should be emphasized that many routes lead to nanotechnology and the field does not encompass a single approach. On the one hand, starting with well known techniques such as devising microelectronics with ever-smaller features may be be part of the

Effective response requires:

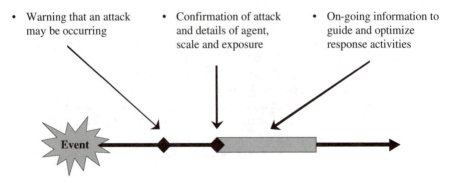

Figure 5.1   Staging of response.

Nanoenabled systems have the potential to:

- Increase speed of measurement
- Improve accuracy
- Improve affordability, deployability, and ease of use of sensor capabilities

Thereby potentially delivering:

- Faster and more certain detection
- Quicker access to higher quality information
- Broader availability of information needed to guide response activities

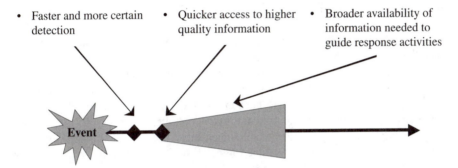

Figure 5.2 Benefits of using nanotechnology to counter threats.

development process. However, recent innovations such as working with deoxyribonucleic acids (DNAs) represent major steps in using nanotechnology to extend the boundaries of conceptual design and laboratory experimentation related to biodefense sensors. Nanotechnology can assess and manipulate molecules, but it involves more than molecular components.

Nanotechnology as applied to defense against biological and chemical agents relevant to human health and agriculture involves various types of components with features measured on a scale of fewer than 100 nanometers (nm). However, the development of specific systems in the field of biodefense equipment is not restricted only to components of this very small size. Integration of detectors, support structures, and electronics will generally involve components across various size scales. In particular, the direct use of microelectromechanical systems (MEMS) and the adaptation of similar systems at nanoscale will often be a feature of approaches to biodefense. Nano-enabled microsystems and macrosystems must be included in a discussion of the use of nanotechnology for biodefense.

Nanofabrication of parts will have to obey many of the principles common to larger system manufacturing, such as providing power and communication channels within the system, but will also have to accommodate the advantages and disadvantages inherent in the development of small components or components employing biological mechanisms. Potential advantages of nanotechnology include (1) the selectivity of molecular recognition, (2) reduced thresholds for detection sensitivity including efforts aimed at single-molecule detection in some cases, (3) the use of living systems and the functions they introduce to sensing capabilities (e.g., biomolecules, subcellular structures, or entire cells), (4) the opportunity to design architectures that will emplace large numbers and multiple types of detectors in a single small system, and (5) the introduction of novel types of detectors including those that can be closely integrated with functions other than sensing, e.g., computational components intended to evaluate the environment and alert users when specific hazards are recognized. The total system package must provide a reliable capability for connecting the miniature components and subsystems with the macroscopic environment and ultimately control by the user. This will sometimes include the need to develop unique displays to accommodate rapid interaction requirements, for example, in telemedicine.

Many applications exist for the variety of sensors under development and also for materials tailored at the nanoscale. Clothing and fabrics with embedded nanoscale materials can introduce barriers and provide for neutralization of toxic agents. Medical sensors in conjunction with substrates placed inside the body can lead to rapid diagnosis and therapeutic opportunities. Structures can be designed to be multifunctional. "Smart" materials that combine the functions of sensors with unique material properties provide examples of the power of integration of nanocomponents.

The implementation of nanotechnologies must accommodate many practical considerations. Disadvantages include the needs to (1) interconnect large numbers of disparate parts, (2) provide a cost-effective mass manufacturing technique for unique nanocomponents, (3) attend to the special needs of living organisms or biological materials, (4) assure redundant designs to compensate for errors in very small and unique components, and (5) provide a path from operating under laboratory conditions to simple and reliable use in the field by nonspecialists. Large-scale production methods for nanomaterials and the subsequent use of the products must not contribute to environmental damage or introduce health problems (e.g., a recent controversy relates to the potential risk of invasion of the body by very small particles). International cooperation is seen in many research efforts, and as products become widely available, decisions will have to be made to assure that mutually acceptable uses and industrial procedures are in place where applicable. Additionally, such cooperation is required to facilitate the employment of biodefenses throughout the world whenever a need may arise. The future development of nanotechnology-based biodefenses will take place along many technological routes, and some of these may require unique guidelines.

Figure 5.3 highlights the application of nanotechnology to biodefense and against various classes of threats amenable to similar technological fixes. A major challenge is to identify and develop a suite of potential nanobiosensors suitable for detection, classification, and alert. In some cases this can be done with samples brought to a laboratory. Important classes of threats — pathogenic organisms introduced naturally or deliberately, certain types of biotoxins, chemical agents intended to directly affect people, and hazardous toxic materials found in the environment or even in food supplies — often require that this challenge be met rapidly by the user *in situ*. This provides an important motivation to search for instruments that are portable, can withstand a range of environmental conditions including moisture, temperature and dirt, and deliver results without additional off-site laboratory analysis. It is also important to increase our understanding of materials and delivery methods that could

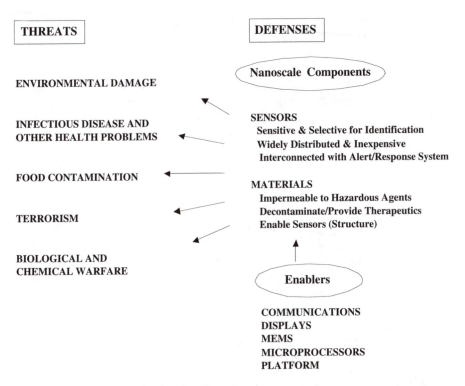

Figure 5.3 Combating major threats with nanoscale components.

protect or remediate threats. Other enabling components of a defense system may find solutions in nanotechnology as R&D progresses. For example, nanoelectromechanical systems (NEMS) are under study and many efforts are underway to reduce computer components farther down the nanoscale.

We will explore many reported examples of components and systems now in research and development and that may lead to applications in the years ahead. It is clear that the development of nanotechnology relevant to biodefense is supported by the efforts of numerous groups throughout the world and includes multinational research. This suggests that the time is ripe to give further consideration to global interactions that will improve our ability to respond to future biodefense conditions. An understanding of how to proceed starts with exploration of opportunities arising from research laboratories.

II. SENSOR COMPONENT OF NANO-ENABLED BIODEFENSE

A. Importance of Information in Biodefense and Sensor Characteristics

In order to address potential threats posed by biological and chemical agents, a key element in any biodefense strategy is the capability to gather information. Both biological and chemical agents cause harm when individuals or areas are exposed to their effects. The potential scale of an event increases as an agent spreads, whether via a chemical, biotoxin, or other noncontagious bioagent (such as anthrax) release, or whether an attack or natural outbreak involves a contagious disease. Real-time information about (1) the initial occurrence, (2) the nature of the agent or agents involved, (3) the individuals exposed, (4) the geographic spread, and (5) the source of the agent or agents can allow military or response organizations to act quickly to both treat the initial victims and minimize the overall effect on the local area, region, or nation.

Similarly, real-time information is also critical in protecting individuals — responders or soldiers — involved in biodefense activities. Because of their roles in responding to events, all types of responders are at much greater risk of exposure to harmful agents than the general public. Beyond the need to protect responders for their own sakes, it is also critical to preserve the abilities of responders and their organizations to act as an event evolves. If the force protection needs of individuals involved in countering such events are not considered, the early phases of a biological or chemical attack may seriously damage the response capabilities of an entire nation.

Individuals need information about their own environments and exposures to guide protective actions and support effective operational and tactical decision making. Complete and timely information will also allow the responders not to over-respond. Protective suits and other equipment are burdensome. Improved sensors may prevent responders from needlessly donning disruptive protective gear. Also, improved information may prevent unnecessary treatment of uninfected individuals. Treatments can have deleterious side effects so the elimination of unneeded treatment may also save lives.

The design of sensor and information gathering systems to provide the information needed to support biodefense activities involves five key technological considerations:

Speed — Depending on the threat involved, biological and chemical agents have the potential to act or spread rapidly. The faster information is made available, the more valuable it is.

Accuracy — Because of the potential stakes involved, the accuracy of biodefense sensor systems is critical. Missing the release of a biological or chemical agent could result in casualties and costs that would be preventable with accurate information. Conversely, false alarms have significant costs as well. Triggering false response actions imposes financial costs and can seriously jeopardize trust in detection systems and in the public officials who use the information.

Ease of deployment — In order to provide the protection needed for individuals, sensor systems that can provide information in the field are superior to those that cannot. Many techniques can identify biological and chemical agents when samples are brought into a laboratory setting. In addition to slowing the availability of needed data, the intermediate steps of laboratory testing also make information less accessible for operational decision making in responding to these events.

Affordability — Ideally, sensor systems should be broadly available and deployed to provide information about wide areas and to large numbers of individuals in the field. Because of practical resource constraints, reducing the costs of individual sensor and detection systems can greatly facilitate this goal.

Ease of use — Because of the wide variety of individuals who need access to biodefense information, systems that are easy to use and do not require significant operator training or intervention are preferred. Ideally, such systems should be passive and their methods of use obvious to individual users. Results should also be easy to interpret.

Because of the potential utility of nanotechnology in the sensing field, recent advances in this area have the potential to significantly contribute to improved biodefense capabilities. Increasing control at the nanoscale level produces opportunities to develop smaller, more specific, power-efficient, and cheaper sensors for chemical and biological agents. The miniaturization of these capabilities increases their potential utility in sensor network applications and has begun to make it possible to integrate them into equipment easily taken into the field by individuals involved in biodefense. Table 5.1 summarizes the general sensor categories discussed in this chapter.

III. NANO-ENABLED SENSORS FOR MONITORING EXPOSURES

Mazzola (2003) reviewed several nanotechnology applications for biotechnology and their anticipated application timelines. The earliest products most applicable to biodefense are microfluidics and nanoscale material manipulations for making sensors. In addition to protecting against threats from agents known to have potential in biological warfare (BW) and terrorism, sensors will have applications to civil problems, for example in providing alerts and characterizing environmental contamination.

These microfluidic and early manipulation products will be followed by composite materials (peptide–lipid assemblies and fabrics) and biosensors (carbon nanotube arrays). Nanotechnology for drug delivery and tissue engineering appears several years from use but is approaching the clinical testing phase today. Nanotechnology that includes the integration of nanoelectric devices such as implantable sensors combined with response systems for drug delivery has the longest timeline to development.

Other sensor technologies are being applied to environmental monitoring for defense against terrorist attacks on water and agricultural supplies or products. For example, Sensicore reports the use of a polyurethane-based sensor membrane with ion-specific binding channels for electrolytes (Yoon et al., 2000) for basic water quality indicators (chlorine, pH, alkalinity, conductivity) and amperometric sensors utilizing permeable polymers for the detection of trace metals such as arsenic, lead, and mercury. The goal is to produce a field-usable device.

Doranz (2003) reported the application by Integral Molecular, Inc., of lipoparticle technology for the detection of biodefense pathogens. Lipoparticles are nanometer-scale spheres surrounded by lipid bilayers embedded with conformally intact integral membrane protein receptors of interest. Bindings to the receptors are then read via an optical biosensor mechanism (see Hoffman et al., 2000).*

---

* See also http://www.integralmolecular.com/lipoparticles.htm (last accessed 12/3/03).

Table 5.1   General Categories of Nano-Enabled Sensors

Category	Example Technologies
Airborne Exposures	
Nanostructured Films	Nanostructure oxides on sensing surface
Combinatorial Microarrays	Nanoscale components in MEMS arrays
Resonant Mass Sensors	Measurement of resonant frequency shifts due to molecular absorption
Biosensors	Binding antibodies to fiberoptics
Electronic Noses	Amplifying chromophore quenching; Polymeric thin films; Gold nanoclusters; Surface acoustic waves
Contact Exposures	
Microcantilevers for Biosensing	Microcantilevers bound with biological analytes causing displacement
Nanoparticles and Nanocrystals	Functionalized with complementary oligonucleotides and protein analytes
Functionalized Nanotubes and Nanowires	Functionalized DNA or protein coatings and built-in detection mechanisms
Nanoscale Components of Sensing Systems	Nanolasers for detection of assay light absorption or emission; Nanoscale, thermally switchable polymer film in a microfluidics device; Nanolithography of biological molecules and sensing materials; Nanoparticle arrays on surfaces; Functional 3D nanostructures; Molecular imprinting
Modified Biosystems	
Phage Display	Engineered protein binding sites on outer surface of phages or viruses
Whole-Cell Sensing Systems	Engineered alarm systems: bioluminescent genes or colony death
Nonbiological Sensor Materials	
Fibers, Fabrics, Membranes, and Textiles	Breathing clothing that prevents external liquids and aerosols to enter; Biocides and chemical catalysts in materials and clothing; E-textile circuits for sensors, processors, and actuators
Conducting Polymers	CB agent sensors woven into fibers; Nanotubes
Nanoscale Decontaminants	Nanoscale decontamination particles; Nanoscale entrapment materials

A. Nano-Enabled Sensors for Monitoring Airborne Exposures

This section will provide a generalized discussion of gas phase sensors and their many applications. In a report of the American Vacuum Society (2002), Baker et al. reviewed the use of nanostructured films for gas adsorption, desorption, or reaction to generate measurable changes in electrical conductances. Particle size and surface structure affect the chemisorption and electrical properties of films, and various nanostructured oxides must be deposited on the sensing surface (Panchapakesan et al., 2001). Microheating a sensor surface increases the performance of the sensor (Semancik et. al., 2001). Other sensor effects are anticipated through the use of different nanostructured materials. Combinatorial microarray methods were employed by the U.S. Department of Energy for hazardous waste detection. The

U.S. Defense Threat Reduction Agency has demonstrated the use of such microsensors to detect chemical warfare agent simulants.

One general class of sensors based on mass measurements for chemical vapors has been suggested in terms of frequency encoding for use with resonant mass sensors (Guan, 2003). It makes use of frequency changes caused by the absorption of molecules on a sensor surface, and the subsequent measurement and application of a Fourier transform method to detect the frequency shift. One objective is to miniaturize the electronics and reduce the size of instrumentation in an effort to make the equipment portable. Such methods have been demonstrated for multisensor detectors. Biosensors serve as natural conduits for detecting bioagents and provide for biodefense detection and identification needs. Nanotechnology can tailor and miniaturize biosensors and furnish additional phenomenology and sensor architectures for detecting threats.

1. Biosensors

The use of organisms, subcellular structures, and biomolecules in sensors is well known. Many types of biological materials or molecules can act as receptor components, e.g., antibodies. These must be combined with detectors such as gas-sensing electrodes. A detector may be miniaturized but remain far from being a nanotechnology component even though its scale has been reduced. It is often advantageous to use a fluid environment for supporting these components in a detection system, and many examples will be discussed in the section on monitoring contact exposures. Biosensors may also apply to gas phase detection.

An example is the piezoelectric immunosensor. This device measures changes in mass on a crystal surface by measuring changes in the resonant frequency of a piezoelectric crystal (Kumar, 2000). The change is proportional to the sampling time and the concentration of analyte in a fluid flowing past the surface. A highly selective receptor is required. Kumar described the application of this class of sensor for detecting tuberculosis and other mycobacterial antigens.

It is often valuable for human health considerations to minimize detection and identification times. Development of real-time biosensors could also aid in responses required to characterize a terrorist attack on populations, agriculture, or food supplies, and minimize the harm done. In the case of a waste site, sensors can be employed at borders and neighboring sites or on samples retrieved from these locations to help manage outflows of effluents (or groundwater for contact sensors such as those discussed in the following section).

An operational issue for any biodefense sensor system is that even inexpensive sensors have practical limitations on what sites and areas they can cover if they must be widely distributed to be effective. In particular, caution must be exercised to recognize that small size, reasonable cost, and detection sensitivity do not necessarily mean that an impregnable field of sensors can be dispersed in an area to catch a potential threat. The gas throughput and diffusion of agents to the volume actually sampled may impose practical limits on detectability.

Many pressing problems could benefit from the continued development of biosensors. Future challenges will include the need to rapidly identify dangerous toxins or organisms in backgrounds of organic materials. The outbreak of diseases caused by prions among cattle and humans in recent years, for example, bovine spongiform encephalopathy (BSE) and transmissible spongiform encephalopathies (TSEs), respectively, may lend itself to prevention and detection using a combination of biotechnology and nanotechnology. Prion diseases are difficult to diagnose other than by analyzing symptoms or performing postmortem analysis. Perhaps some molecular-based detection methods can be developed to detect these relatively small biomolecules (although this example may require a fluid and not a vapor detection method). A key question for the future use of nanotechnology is whether the exploitation of molecular selectivity can protect a food supply from contamination by particularly difficult or currently impossible-to-detect toxins? Biosensor technologies will be generally more useful for organisms that are difficult to culture.

2. Electronic Nose

The artificial or electronic nose now under development is intended to serve the same purpose as a canine (or other animal) in the service of detecting explosives or other classes of dangerous chemicals (Yinon, 2003). This technology will lead to safer methods of searching for landmines, terrorist bombs, drugs, accidental industrial chemical releases, and the presence of any hazardous chemical. The use of microtechnology and nanotechnology (if cost-effective) will allow the continued miniaturization and widespread distribution of this class of detector beyond traditional uses with the deployment of mobile detectors such as mass spectrometers or gas chromatographs. Yinon describes examples upon which electronic noses may be based. In addition to MEMS, he examines four nanotechnology-based approaches: amplifying chromophore quenching, polymeric thin films, gold nanoclusters, and surface acoustic waves (SAWs). Other methods may also be developed.

The purpose of the quenching detector is to increase sensitivity to reactions with specific classes of vapor molecules by causing the absorption of a single photon to establish a chain reaction in a sequence of chromophores. Trinitrotoluene (TNT) and other nitrocompounds are cited as examples of molecules that can be sensed, and this clearly has application to the problems of detecting landmines and unexploded ordnance. As Yinon notes, the United Nations is concerned with the enormous number of landmines left in many countries.

Yinon refers to a class of thin films that use changes in resistance to detect molecules absorbed in a set of polymers (Lewis, 1995). It is of interest that this method uses multiple sensors and neural net analyses to identify the vapors. The development of hardware is not the only important effort needed to make nanotechnology work well. The development of software is also vital, especially because numerous sensors and sensor types may have to act together to discriminate and measure contamination. Cyrano Sciences, Inc. of Pasadena, California, developed commercial electronic noses based upon composite polymer–carbon nanoparticle films that are currently being tested in biodefense applications (www.cyranosciences.com).

The nanocluster devices described by Yinon also make use of electrical properties like resistance and conductivity. The gold particles used in the vapor detection devices are described as smaller than 5 nm in diameter and surrounded by single-layer organic shells. Importantly, they were found to be capable of detecting many classes of vapors including those of a number of explosives.

SAW devices use a piezoelectric substance upon which an acoustic wave is generated by an electric field, travels along the surface, and is affected by any vapors present. This interaction may be detected and analyzed. These types of detectors have been the subjects of numerous studies and may be used to search for explosives and hazardous materials.

Yinon concludes that the many types of detectors now in development may apply across a broad set of problems. The threat of terrorism with conventional weapons makes the development of small, inexpensive explosives detectors that can be mass-produced of high interest. The same classes of sensors that protect against chemical and biological agents may be used for detecting explosive molecules. Protection of aircraft against smuggled bombs is another clear example where detectors designed for picking up explosives are of interest. An important question for further consideration is whether detectors at checkpoints could reduce the incidence of suicide bombers who proceed through public areas. The manner in which the sensors are deployed as well as their technical limitations must be considered for feasibility of use in any scenario.

The detection and identification of pathogens and biological agents in general can also be performed with an electronic nose, as indicated by the previous discussion of biosensors. This has been demonstrated with discrimination among anaerobic bacteria grown *in vitro* (Pavlou et al., 2002). Multisensor arrays containing conducting polymers were used to detect gases. Again, this detection method employed neural nets. Other techniques are, of course, also applicable to detecting organisms and may have their own advantages (e.g., mass spectrometry can be used to identify species).

Conducting polymers have been used to discriminate wine aromas (Guadarrama et al., 2001) and this example suggests nanosensors may eventually lead to food characterization — i.e., determining when a food has gone bad. Preservation is a more ambitious goal and requires sensing to detect processes before spoilage.

MEMS-based noses use cantilevers that respond when exposed to gaseous analytes. One such array uses the swelling of one or more polymer coatings as an indicator for detection using optical measurements of deflection (Baller et al., 2000). Experiments with these devices investigated the detection of organic molecules including methanol, toluene, and ethanol and will be further discussed later.

Biological threats and inorganic toxins (or explosives in the case of terrorism) are not the only hazards that may be detected with devices that operate at the nanoscale. It is also possible to detect the presence of radioactive materials with cantilevers that are kept at a distance of a few nanometers from an insulated metal surface (Thundat and Brown, 2002). Alpha particles can be detected as a result of charge accumulation or frequency shift. This method can be very sensitive. Single alpha particles have been detected. The time required for detection can be improved with the use of large-area detectors. The technique could be useful in detecting many types of low level activity by sampling gas, but alpha particles have very short ranges and will not penetrate far, so the method will not detect radioactivity that is well contained or shielded by soil or other factors. Vapors from a suspected contaminated material are needed or a detector must be demonstrated to be capable of measuring other, more penetrating particles such as gamma rays.

The sensitive detection of low level radioactive contamination is of interest. Some hazardous sites contain both chemical toxins or biologically hazardous materials and radioactive wastes. In these cases, it is important to be able to detect radiation while sensing molecules. Examples include remediation and attempts at containment of affected areas resulting from activities of nuclear power industries, research laboratories, weapons programs, and from radioactive medical waste. Further descriptions of using arrays of cantilevers for sensing biomolecules through the use of liquid contact media appear in the next section.

B. Nano-Enabled Sensors for Monitoring Contact Exposures

In nano-enabled sensing technologies, the key concept in sensor design is taking advantage of biological or chemical binding, specific recognition, reactivity, or other mechanisms to access desired information. Sometimes these sensor systems employ actual biological systems or materials, sometimes designed analogs or substitutes. The role of nanotechnology in the system is to gather the information; build the detection, signal transduction, or coupling mechanisms to convert the biological or molecular event into a detectable signal; and make the systems practical and rugged enough to fulfill their missions.

This section will review different nano-enabled approaches to building sensing systems relevant to biodefense applications. Because of the rapid evolution of and advances in this research area, the applications discussed and examples cited focus on the recent literature. Both in the interest of brevity and due to the sheer scope of the topic, it was not possible to comprehensively review even individual technology strategies. As a result, the technologies and applications cited should be viewed as promising examples, selectively drawn from the large body of quality work on relevant sensor technologies. In contrast to the previous discussion focusing on environmental and gas phase monitoring, this section focuses on contact sensors — technologies frequently applied to samples in solution and applicable to providing information on individual exposures or potential threat agents in an individual's immediate environment.*

One example that is relevant to discussions about protecting health and sampling the environment is the use of fiberoptics in biosensors that may be classified as nanotechnology-enabled devices. The fiberoptic components are clearly not examples of nanotechnology. Only the fact that enabling molecules are introduced for the detection process makes fiberoptic technology relevant to discussions of nanotechnological devices.**

---

* It should be noted that the distinction drawn here between individual and environmental sensor systems is not entirely a clean one. In many cases, sensors that would be applicable to measuring the presence of biological or chemical agents in ambient air (an environmental application for purposes of this discussion) could be applied to an individual's monitoring needs as well in a somewhat different technological application. Similarly, many of the technologies that we consider individual (e.g., nanowire sensors) could also be applied in environmental applications. The distinction was drawn mainly for organizational reasons and, while imperfect, is useful for that purpose.

** One potential challenge in understanding nanotechnological biodefense is drawing the distinction of what is accepted as nanotechnology. However, for purposes of operational usefulness of devices that are produced this will often be a matter of semantics.

One method for exploiting fiberoptics is to bind antibodies to the fiber, allow antigens to be captured, introduce antibodies with fluorophores, and sample these with a laser reflected through the fiber (Lim, 2003). This method may be useful for environmental monitoring for pathogens because a highly sensitive and specific method reduces the time required to cultivate, separate, and identify organisms and other biological agents. Lim reports that very good sensitivity and selectivity can be achieved, for example, detection of *Bacillus anthracis* at 10^5 colony-forming units/ml, cholera toxin at 100 pg/ml, and TNT at 10 ng/ml. In addition to high sensitivity, Lim points out that biosensors can reduce the time required to detect these hazards from hours or days to minutes. This capability would enable detection of contaminants in the field.

Binding of fluorophore molecules (placing functional groups on a structure) is an example of a type of approach to nanotechnology. Investigations of the properties of such molecules as DNA and proteins further demonstrate the connection of biotechnology and nanotechnology. Biological structures and molecules as detection devices can be integrated with many devices other than fiberoptics. For example, microbeams can be fabricated with nanostructured surfaces that detect proteins (Dutta, 2342). Microbeams and additional sensor enablers will be discussed in the following sections.

In pursuit of highly miniaturized sensors for detecting biological molecules and chemical agents at very low concentrations, a number of nanoscale detection techniques have the potential to achieve nearly single-molecule detection. One family of technologies includes an entire sensor built at the nanoscale. These technologies have as their bases individual nanostructures such as nanotubes, nanowires, nanoparticles, or microcantilever systems.

1. *Microcantilevers for Biosensing*

Scientists have been using micromachined cantilevers as force probes in atomic force microscopy (AFM) for several years. The extreme sensitivity of the probes has prompted research into incorporating these structures into biosensors. Recent studies report success in detecting the binding of biological analytes on the surfaces of the microcantilevers that cause tiny conformational changes within the microcantilever structure. Depending on the type of analyte for which the device is designed, the magnitude of the conformational change, and the sensitivity and specificity of the microcantilever structure, these systems are useful in various kinds of biosensors that may be critical tools in biodefense efforts.

Microcantilevers are tiny plates or leaf structures, usually measuring 0.2 to 1.0 µm thick, 20 to 100 µm wide, and 100 to 500 µm long. One end is connected to a support. In order to make them useful as biosensors, the plates, usually made of silicon, are coated on one side with a different material that can then be functionalized in various ways. Ideally, when exposed to the analyte of interest, the functionalized side will undergo stress-induced conformational changes while the other side remains unaffected, resulting in plate deflection.

Three mechanisms of analyte-induced stresses have been described (Sepaniak et al., 2002): compressive stresses caused by the physical expansion of the functionalized side due to analyte binding, swelling of a thin film due to analyte adsorption, and expansion of the functionalized surface due to interstitial forces caused by analyte binding. The mechanism of the sensor depends on the type of responsive coating used in forming the cantilever. There are also several options in the type of analyte the sensor is designed to detect. Studies to date have used microcantilevers to detect volatile organic compounds (VOCs), ionic species, proteins, and oligonucleotides. In most cases, detection ultimately involves the conversion of plate displacement into electrical signals via the projection of a laser beam onto a position-sensitive photodetector.

Of great importance in biodefense are sensors that can detect proteins of interest amid varied background environments. Protein sensors based on microcantilever technology take advantage of structural changes and/or changes in the net or conformational charges that result from protein binding. The unique interaction of each protein with the functionalized surface of the cantilever results in differences in surface stress that can then be measured through the magnitude and direction of plate deflection. This technique has been demonstrated successfully in several cases including the detection and differentiation of low density lipoproteins (LDLs) and oxidized LDLs (oxLDLs) (Moulin et al., 2000) and two forms of prostate-specific antigen (PSA) in humans (Wu et al., 2001). Similar studies cover functionalized microcantilevers with oligonucleotides complementary to the DNA or RNA of interest (Hansen et al., 2001).

Specificity concerns have been addressed by detecting protein analytes amid other ambient molecules, mimicking, to some extent, the proteins' natural environments (Moulin et al., 2000). The sensitivity of the biosensor also improves as a function of its small size. Detection of analyte has been achieved at concentrations as low as 0.2 ng/ml (Wu et al., 2001). This is another example of the high sensitivity of nanotechnologies and is several-fold more sensitive than technologies currently in use. Deflection responses up to several hundred nanometers have been shown to vary linearly with analyte concentration (Tipple et al., 2002).

Nucleotide mismatches can be detected at the level of a single base pair, with deflection increasing predictably with the number of mismatched base pairs (Hansen et al., 2001). The false positive rates should be estimated for any device developed. This is important for instruments that are extremely sensitive to low detection thresholds. The magnitudes of false positives cannot be declared for these examples at the current time because few of these technologies have been tested under conditions that would provide realistic estimates of rates.

2. Nanoparticles and Nanocrystals

Similar to microcantilevers, nanoparticles and nanocrystals can also be functionalized and integrated into biosensor systems. The nanoparticles are usually made of gold and have diameters smaller than 70 nm. Several mechanisms for the detection of complementary oligonucleotides and protein analytes have been developed over the past 5 years, although most have focused on DNA and RNA detection. In addition to the functionalized or sticky oligonucleotides that are complementary to the target analyte, the particles must be detectable and/or distinguishable in some way.

Recent studies have taken advantage of spectroscopic techniques including the control of fluorescence intensity (Dubertret et al., 2001) and spectrum fingerprinting (Cao et al., 2002; Gerion et al., 2002). Other approaches have involved measuring electrochemical changes resulting from hybridization or enzyme activity (Park et al., 2002; Chakrabarti and Klibanov, 2003; Xiao et al., 2003). In each case, the binding of the complementary molecule induces a measurable change in the nanoparticle that is then detected.

Using nanoparticles or nanocrystals in biosensors can enhance the speed, portability, sensitivity, and selectivity of the process. For instance, the need for a laboratory to process samples through the sensor may be removed with the development of dry reagent systems (Glynou et al., 2003). The instability that often plagues DNA hybridization techniques can be alleviated through the use of DNA analogues (Chakrabarti and Klibanov, 2003). Unlike the microcantilever systems, this type of sensor is not often used to detect or measure the extent of nucleotide base mispairings; however, in some cases, this level of sensitivity is achieved through techniques that exploit the decreasing stability of imperfectly hybridized oligonucleotides (Dubertret et al., 2001; Park et al., 2002).

3. Functionalized Nanotubes and Nanowires

Other recently developed tools useful in the miniaturization of biosensing devices include nanowires and nanotubes. Similar to both microcantilevers and nanoparticles, nanotubes and nanowires can be integrated into a biosensor through functionalized coatings and built-in detection mechanisms. Nanotubes are particularly promising because of their durability and extreme sensitivity to electronic transport and voltage caused by interjunction temperature differences (Baughman et al., 2002). Their shape and durability make them ideal candidates for integration into portable self-contained sensing devices (chips or other immobilized arrays) that will perhaps make on-site applications away from the laboratory possible.

These qualities have been used in several studies demonstrating successful biosensors for DNA, proteins, and enzymes. It has been shown that DNA (Williams et al., 2002) and proteins (Besteman et al., 2003; Star et al., 2003) retain their biological activities when covalently bound to nanotubes. Subsequent analyte binding to these active molecules results in detectable changes in the nanotubes that can be measured at extremely high sensitivity, in some cases even allowing the detection of a single-molecule redox reaction (Besteman et al., 2003). The tiny electrochemical or physical changes on the surface of the tube or wire resulting from the biological reaction of interest cause measurable changes in its conductance or resistance that can then be easily detected and quantified (Kong et al., 2000; Besteman et al., 2003). Nanotube-based DNA sensors can be made even more stable when DNA analogues are used (Williams et al., 2002) and protein sensors gain even greater specificity when shorter nanotubes are used (Besteman et al., 2003).

Mere detection of a biological material is not sufficient. One issue with detection technologies is to determine whether the organism is alive so that a terrorist's use of dead anthrax spores, for example, does not trigger a response that gives a false indication of the threat. Depending on the sensing mechanism, sensors at the molecular level may not be able to detect whether an organism is living or dead, but simply whether a certain protein structure or molecule is or is not present. This limitation must be recognized when planning how to use sensors. The next section will provide further technical reviews of several classes of nanoscale components for sensors.

IV. NANOSCALE COMPONENTS OF SENSING SYSTEMS

In a laboratory context, a wide variety of techniques have been developed for the detection of biological and chemical agents. These larger-scale sensing approaches frequently utilize techniques such as light absorption, emission, radioactive tags, electrochemical detection, chemical analysis, piezoelectric, micromechanical methods, and other techniques to directly detect biological or chemical events at the molecular scale.

Although these techniques provide ways to assay samples for particular biological threats or chemical agents, the need for laboratory instrumentation and the time required to perform detailed analyses make them less than ideal for biodefense applications. While such approaches may provide accurate ways to characterize an unknown threat, they cannot deliver the speed, ease of deployment, cost, and ease of use that would be most useful in biodefense applications.

Significant progress has been made in developing alternate detection methods for these assays or deployable versions of such laboratory-based approaches to improve their potential application to biodefense. The capabilities provided by nanotechnology — such as designing highly miniaturized structures, building novel sensor components, or new detection strategies — can make significant contributions to that effort. Because of the variety of technologies involved in manufacturing sensor systems, any discussion of the contributions of nanotechnology in this area is by definition incomplete.

To demonstrate the varied contributions nanotechnology could make to such detection systems, it is instructive to consider two disparate examples. Building on current research in nanoscale materials and physics, nanoscale lasers could provide light sources for miniaturized versions of an assay based on detection of light absorption or emission. In a very different technology area, significant efforts in biological and analytical chemistry have been devoted to developing chip-based detection methods.

Significant progress has been made developing microfluidic devices that can assay for a range of biological or chemical agents of interest. Recent examples include assays for DNA analysis (Breadmore et al., 2003), explosives (Wang, J. et al., 2002a), nerve agents (Wang, J. et al., 2002b), and bacterial spores (Stratis-Cullum et al., 2003). A central goal of such research is the eventual integration of many different assay methods in a single device — a "lab on a chip" that could provide a user with a wide variety of information on all known threat agents. Such chip-based technologies could have a large number of components that nanotechnology or nanoscale manipulation could significantly improve. For example, a recent report by Huber et al. (2003) described development of a nanoscale, thermally switchable polymer film. In one state, the film readily absorbs proteins; in the other, it repels them. Such a component could be useful in a range of microfluidic applications for protein concentration or purification before analysis.

Rather than seek to survey nanoscale advances in all fields potentially relevant to detection system design, we have chosen to focus this discussion on four major areas. The subsequent sections will examine recent examples of research in nano-lithography of biological molecules and sensing materials, assembly of nanoparticle arrays, construction of functional nanostructures, and design of individual recognition elements at the nanoscale through molecular imprinting. The selection of these areas continues the approach adopted in the previous section on individual nano-structures by focusing on construction of nanoscale structures in design and development of novel sensors.

A. Nanolithography of Biological Molecules and Sensing Materials

The positioning and immobilization of biological molecules or sensing materials on two-dimensional surfaces can provide the starting point for development of a range of different sensing and assay methods. For example, for many years the recognition of antibodies — proteins used by the immune system to recognize and defend against external threats — for their binding targets has been used in assay design. When these assays are performed connected to a surface or other solid support, the binding of the target molecules can be detected by a range of techniques. Similar surface-based detection methods can be used to detect specific sequences of DNA diagnostic for particular bacteria or viruses that pose biological threats, small molecules including chemical agents, and other substances.

Although such surface-based assays can be carried out for a single protein or analyte of interest, performing assays with arrays of different molecules creates the potential to gain significantly more information from a single detection device. Creating arrays of proteins at the nanometer scale can make it possible to examine a wide variety of components of a complex mixture or environmental sample in a highly miniaturized device. In an alternate application, arrays of a single protein can also be used to perform many replications of the same assay on a single chip.

The technique of dip-pen nanolithography (DPN) makes it possible to create nanoscale structures on a surface. The technique involves the transfer of the molecule of interest using a coated AFM tip. Depending on the particular application, DPN can be applied in a variety of ways to construct nanoscale structures on surfaces. To build a surface array of a single protein, Hyun and colleagues (2002) used DPN to create an array of protein dots with feature sizes on the order of 230 nm. The stepwise technique they developed relied on the recognition properties of biotin–streptavidin, a pair of molecules that bind very tightly to one another. DPN was used to functionalize a gold surface with an organic linker to which biotin could be connected. Streptavidin could then be used to link the modified surface to biotin-modified proteins. Because many proteins of interest can be biotin-modified, the technique allows preparation of nanoscale arrays from a variety of proteins. Lee et al. (2002) have shown that analogous arrays with feature sizes as small as 100 nm can be constructed by the assembly of proteins onto nanopatterned monolayers on surfaces.

DPN has also been used to construct patterns of DNA directly onto gold surfaces and derivatized silica surfaces. Pieces of DNA derivatized with linker molecules that allow connection to the surface are directly applied with an appropriately selected AFM tip. This process allowed production of surface features on the order of 50 nm in size and the DNA applied to the surface retained its recognition properties for other nucleic acids. DNA applied to a surface can be used to assemble gold nanoparticles functionalized with a complementary strand of DNA (as discussed above). The fact that the method can be used to connect DNA to both metallic and semi-conductor surfaces makes it applicable to a wider range of potential applications.

Beyond using biological molecules as inks for DPN, this technique can also be used to pattern inorganic sensing molecules on surface substrates. Su et al. (2003) demonstrated that approximately 32-nm thick metal oxides can be deposited on prefabricated electrodes as sensor elements. Using the technique, a sensor array with eight elements made from tin oxide doped with various other metals was prepared.

B. Nanoparticle Arrays on Surfaces

The unique electronic, optical, and catalytic properties of nanoparticles driven by their small dimensions make them useful in sensing applications. Building on the advances in individual nanoparticle sensors, a second strategy to develop novel sensing elements based on the construction of nanoparticle arrays on surfaces is being actively pursued.

A number of techniques have been developed to pattern nanoparticles of various materials on surfaces (see Shipway et al., 2000). Lithographic methods, printing methods, and templating assemblies of the particles with other biological molecules such as DNA have all been explored. Nanoparticle arrays demonstrating potentially biodefense-relevant sensing mechanisms include small molecule sensing by binding-induced optical or electronic changes, ion-sensitive field effect transistors to detect small molecule binding to receptors synthesized on particle surfaces, and biological reactivity sensing by nanoparticles enzyme conjugates (see Shipway et al., 2000).

A recent example of a nanoparticle array sensor was constructed by Haes and Van Duyne (2002) of triangular silver nanoparticles. Using biotin–streptavidin binding as a model system for protein binding to the array sensor, they demonstrated that protein binding to the array could be detected by optical methods. The potential for these sensors to be combined with the single nanoparticle sensing mechanisms discussed above could lay the groundwork for parallel sensing of many analytes simultaneously.

C. Functional Three-Dimensional Nanostructures

Just as the properties of nanoparticles make arrays or other aggregates of the particles useful for sensing applications, other three-dimensional (3D) nanostructures can also provide the bases for nanoscale sensors. The large surface areas that can be created in three dimensions of even basic nanostructures can provide convenient bases for the design of sensor systems hosting a large number or variety of active binding sites. Beginning from single-walled carbon nanotubes, Novak et al. (2003)

constructed a nanotube matrix as the basis for a sensor for chemical agents. Upon adsorption of an analyte, single-walled carbon nanotubes can exhibit a significant resistance change that provides a readily detectable signal for sensor design. The use of nanotube matrices avoids the variability in properties observed in applications using individual nanotubes, yet maintains the sensitivity advantages of single nanotube devices.

Other examples of 3D nanoscale structures used in sensing include structures produced by electrospinning poly(acrylic acid)–poly(pyrene methanol), a fluorescent polymer used to sense both toxic metal ions and explosive molecules (Wang, X. et al., 2002) and construction of silver nanowire membranes for small molecule sensing using vibrational spectroscopic detection (Tao et al., 2003).

Nanoscale design of 3D structures can also provide strategies for detection methods that do not depend on external spectroscopic or other techniques. For example, a combination of molecules forming a liquid crystal with receptors for a chemical or biological agent of interest can provide a sensor where the presence of the agent can be read visually. In the absence of the agent, the molecules of the liquid crystal are designed to occupy the binding sites of the receptor molecules. When exposed to the agent, the liquid crystal molecules are displaced and this causes a phase change (and therefore a change in appearance) of the liquid crystal (Shah and Abbott, 2001).

Beyond the construction of bulk 3D structures from nanotechnology components to serve as sensor elements, engineering on the nanoscale can be used to design specific molecular structures to recognize molecules of interest. For example, Kasianowicz and co-workers (2001) began with a protein that acts as an ion channel in the bilayer membranes that surround cells. The behavior of such ion channels can be monitored by electrical currents arising from the passage of ions from one side of the membrane to the other. The researchers designed a polymer that can thread through the channel, thereby blocking it and perturbing the electrical signal. By engineering a binding site for an analyte of interest onto this polymer, this simple system can provide detection information. Binding of the analyte to the polymer will affect its ability to thread into the channel and thereby perturb the electrical signal produced by the system. This perturbation provides a detectable signal for the analyte of interest.

Sasaki and co-workers (2002) similarly designed a lipid membrane sensor system at the nanoscale. The system consisted of a lipid molecule containing both a recognition group (resident on the membrane surface) and a fluorescent reporter molecule. The recognition group binds lead ions. The electrostatic repulsions among multiple bound lead ions cause changes in the distribution of the lipids in the membrane that can be detected by perturbation in fluorescence.

D. Molecular Imprinting: Construction of Recognition Elements at Nanoscale

Although recognition elements for specific molecules can be developed by rational synthetic processes, i.e., designing a single binding site through placement of appropriate functional groups or binding pockets to match the structure of the molecule, such a process is laborious and not always straightforward. As a result, alternative strategies for designing these nanoscale binding sites have been developed utilizing molecularly imprinted materials. To produce a binding site through molecular imprinting, the target molecule (or similar analog) is used as a template. A shell of functionalized and cross-linkable polymers is then formed through interactions between the polymers and the template. The polymers are then cross-linked and the template molecule is removed. The polymer shell that remains is then locked in a permanent geometry complementary to the template molecule. By judicious design of the polymer structures, the nanoscale binding site is allowed to design itself by associating around the molecule of interest (see Haupt, 2003 for a more complete discussion).

The general technique of molecular imprinting has been used to produce a variety of nanoscale structures for sensing applications. Imprinted polymer thin films have been produced to bind analytes (Duffy et al., 2002); synthetic host molecules have been produced by molecular imprinting inside dendrimers (Zimmerman et al., 2002); and polymers specific for nerve agents and explosives residues have served as the bases for optical and other detection systems (Arnold et al., 1999). The technique has even been taken to the point where imprinted polymers have been used to produce surfaces that recognize whole bacteria (Das et al., 2003). Imprinting techniques have also been applied to materials other than cross-linkable polymers. For example, proteins have been used to imprint a layer of sugar molecules that were then immobilized by embedding in a solid support. The resulting complex nanostructure surface was shown to selectively bind the proteins used in the imprinting process (Shi et al., 1999).

V. MODIFICATIONS OF NATURAL SYSTEMS ON THE NANOSCALE FOR SENSING

Beyond strategies based on constructing novel individual nanosensors and utilizing nanostructures in other sensing applications, nanotechnology has also allowed modification of natural systems for sensing applications. The tools of molecular biology and biotechnology have enabled the nanoscale construction of structures in bacteria and viruses, making those organisms into sensing elements for agents of interest in biodefense.

A. Phage Display

The phage display techniques developed by researchers for uses in biosensing exploit some of the existing features of phage physiology and life cycles. Bacteria phages or viruses can express proteins containing binding sites specific to particular molecules on their outer surface. By controlling which proteins are expressed, scientists are able to design highly specific molecular probes. Once exposed to a sample, any phage whose surface proteins are specific to peptides or antigens in the sample can be isolated via standard laboratory affinity and separation techniques.

Phage display is similar to the antibody–antigen detection techniques in use for many decades, but it offers a more stress-resistant and longer lasting binding between probe and analyte (Petrenko and Vodyanoy, 2003). The various factors that can affect the use of phages as molecular probes have been described in detail based on experimental and simulation data (Levitan, 1998). The ability to design specific molecular probes based on phage display techniques through the control and modification of these factors has also been explored (Kirkham et al., 1999; Sblattero and Bradbury, 2000).

These and other studies have shown that the detection of specific proteins and antigens is possible at diagnostically significant concentrations through the use of modified phage surface proteins (Petrenko and Vodyanoy, 2003). However, the use of such pin-pointed specificity may be less useful in realistic scenarios in which a sensor is needed to detect an unknown. Phage display sensors, therefore, will probably play a more significant biodefense role in confirmational tasks.

B. Whole-Cell Sensing Systems

Several other biological systems lend themselves for use in biosensors. Most biological systems-based sensors can provide information about the effects of the agents to be detected on living organisms — information that is usually not obtainable via other sensor platforms (Belkin, 2003). The ease with which scientists can manipulate gene expression in most bacterial cells now makes it possible to modify a system to include a built-in alarm (e.g., expression of bioluminescent genes or colony death) that sounds when particular biological functions are interrupted. These types of sensors are typically used for environmental monitoring but they have also recently been applied to the *in vivo* detection of pathogens (Innovative Biosensors Inc., 2003; Rider et al., 2003).

VI. NANOMATERIALS AS ENABLERS FOR BIODEFENSE

Nanomaterials can contribute to biodefense strategies and implementations in a number of distinct functional ways:

1. As barriers to chemical and biological (CB) agents
2. By providing substrates for CB agent sensors
3. By providing functional sensor components for CB agents
4. As means to store and then release decontamination agents
5. As decontamination agents

We review below three classes of materials that can contribute to these functions, sometimes in more than one way at a time: nanofunctional fibers, fabrics, membranes, and textiles; conducting polymers; and nanodecontaminants.

A. Fibers, Fabrics, Membranes, and Textiles

The current state of the art in the development of fabrics that can in principle integrate the aforementioned functions is described in a recent review article (Schreuder-Gibson et al., 2003). Fabric membranes used for clothing allow moisture vapor from the wearer to escape but prevent liquids and aerosols from entering. Biocides and materials that can hydrolyze and detoxify chemical agents have also been attached to polymers that were then spun into fibers from a melt charged to thousands of volts (Schreuder-Gibson et al., 2002). Membranes produced by this electrospinning process have fibers with diameter in the 100-nm to 10-µm range. The development of fabrics from these fibers and membranes that can incorporate chemical catalysts and biocides is an ongoing research effort at the U.S. Army Natick Soldier Center in Massachusetts (Schreuder-Gibson et al., 2003).

Another approach for integrating materials functions is the development of fabrics and textiles that conduct electricity (electrotextiles or e-textiles) and can thus serve as part of a working circuit that includes sensors, processors, and actuators. The development of e-textiles was pioneered by researchers at IBM and MIT, with a focus on applications that integrate computing (Post et al., 2000). These researchers developed several composite fibers and methods to weave them, join them, and fabricate electrical circuits using them. They suggested future work to increase the durability of composite fibers, integrate them with optical fibers, and achieve elec-trospinning of conducting, semiconducting, and insulator fibers into wearable electronic structures.

Electrospinning was in fact used recently to produce a nanofibrous template for growing conducting polymers for biomedical applications (Lin et al., 2003). A coagulation-based spinning process was also used to produce composite carbon nanotube fibers made into supercapacitors and woven into textiles (Dalton et. al., 2003). Another recent effort is incorporation of full electronic functioning into circuits woven into clothing (see Bonderover et al., 2003). A review article (Natarajan et al., 2003) cites the use of e-textiles as thin film transistors on polymer films and as textile-based batteries and solar cells. Natarajan's group also discusses commercial switches that are woven into fabrics (http://www.softswitch.co.uk/) and their own research on multilayered woven fabric-based electrical circuits. Other developmental applications of e-textiles include sensing liners for monitoring the medical conditions of battlefield personnel, pressure-sensitive switches for space suits, and inflatable airbag systems and radar antenna arrays (Cadogan and Shook, 2003).

Two e-textile applications in the early stages of commercialization are especially relevant to this chapter. The Smart Shirt Project at Georgia Institute of Technology developed a wearable fabric containing a single plastic optical fiber that is spirally integrated into the fabric along with an electrical grid, connectors, and processors that allow plug-and-play sensing, monitoring, and information processing (Park and Jayaraman, 2003; Marculescu et al., 2003).

This technology has been commercialized by Sensatex of New York City. The company offers a SmartShirt® that allows measuring and/or monitoring of individual biometric data (heart rate, respiration, body temperature, caloric burn) and provides read-outs via a wristwatch, personal digital assistant (PDA), or voice (www.sensatex.com). Another product, the LifeShirt™ of VivoMetrics of Ventura, California, is a lightweight, vest-type garment with embedded sensors, a two-axis accelerometer, and a single-channel electrocardiograph for continuous ambulatory monitoring of patients, with provision for a patient to enter time- and date-stamped data about symptoms, moods, and activities (www.vivometrics.com). Future generations of such garments that incorporate sensors and processing could provide platforms for nanoscale CB sensors.

B. Conducting Polymers

Conducting polymers constitute a unique class of materials that can provide CB agent sensing capability and allow integration into fibers, fabrics, and membranes. Nanoscale fibers (<100 nm in diameter) of the polyaniline and polypyrrole conducting polymers and their blends with common polymers such as polystyrene and polyethylene oxide have been fabricated using the electrospinning technique described above (MacDiarmid et al., 2001), enabling their incorporation into fabrics and membranes.

Another approach useful for incorporation of conducting polymers into fabrics is the production of nanotubules of these materials that can be synthesized using a nanoporous membrane as a template (Martin, 1994; Parthasarathy and Martin, 1994; Cepak et al., 1997) or via a supramolecular self-assembly process (Qiu et al., 2001; Liu and Wan, 2001). Conducting polymers have been successfully demonstrated to serve effectively both as chemical (Janata and Josowicz, 2003) and biological (Sadik, 1999) sensors via observation of changes in resistance or other electrical properties produced by adsorbate molecules. For example, several of the electronic noses described in Section III.A.2 are based on pattern recognition of electrical signals detected from arrays of conducting polymers exposed to chemical (Barisci et al., 2002) or biological (Pavlou et al., 2002) agents.

C. Nanoscale Decontaminants

A recent workshop held by the American Vacuum Society (AVS) under the auspices of the U.S. National Nanotechnology Initiative (AVS, 2002) described two different nanomaterial approaches to CB agent decontamination: (1) the use of nanoscale particles and (2) the entrapment of nanostructured materials within the interiors of high porosity carrier networks. The nanoscale particle approach takes advantage of the increased surface area and the presence of a higher number of reactive sites at the surfaces of metal oxides known to have reactivities to CB agents.

For example, powders of magnesium oxide and aluminum oxide with diameters of a few nanometers prepared via an aerogel method have been demonstrated both in dry powder and halogenated forms to be highly effective biocides (Koper et al., 2002). Nanoscale magnesium oxide has also been demonstrated to react strongly with chemical warfare agents to detoxify them (Wagner et al., 1999). A recent finding noted in the AVS workshop is that nanoscale powder combinations of magnesium oxide and aluminum oxide enhanced reactivity against chemical warfare agent surrogates, as compared to either powder used alone.

VII. INTEGRATION AND MULTIFUNCTIONAL SYSTEM CONCEPTS

Bringing the sensing, processing, and decontamination or neutralization functions described above together could in principle be achieved within a multifunctional material designed at the nanoscale. Separate elements that would have to be integrated in such an approach can be seen in a variety of recent advances.

Table 5.2 lists the basic system functions as they may be met currently and in the near and far terms for personal biodefense system concepts. The two basic types of biodefense systems are (1) a protection system that prevents exposure in the first place and (2) a sense-and-respond system that detects an attack and initiates appropriate responses to mitigate the damage. Examples of protection system trends foreseeable in the future include nanofabrics that contain increasingly tunable and automatic responses to exposure.

For sense-and-respond systems, nanotechnology appears poised to not only shrink the scales of individual components within systems but also to achieve multifunctional advances to the point where sense and response functions will be integrated. In simpler examples, nanoparticle entrapment and biocide fabrics do not need separate sensor functions; the materials respond as needed. However, future e-textiles could integrate multifunctional components to the point where they are not separate in the traditional sense. For example, conducting polymers or carbon

Table 5.2 Integrated Nano-Enabled Personal Biodefense Systems with Multifunctional Components

	Examples of Personal Protection System Concepts		
	Today	Near Term	Far Term
System Type and Function	Intercommunicating multicomponent systems	Nano-enabled communicating components	Multifunctional integral system
Protection Systems			
Protection	Standard hazmat protection suit	One-way vapor fabrics	Active fabric breathing
Sense-and-Protect Systems			
Sensor	MEMS-based biosensor	Molecular-based sensor module	Nanoparticle entrapment; Auto-response biocide fabrics; Conducting polymer or carbon nanotube sensor integrated with microneedle drug delivery on same fabric and bio-energy recovery with fiber-based batteries
Internal system communication	Short-range wireless	e-Textile	
Processing	Microchip	e-Textile processors	
Response	Polymeric-based drug time release	RF-signaled polymeric-based drug release	
Power	Macro-scale Li ion batteries	Thin film Li ion microbatteries	
Activation	Alarm with human confirmation	e-Textile alarm and activation	Automatic response
Substrate	Multicomponent wireless system	Multifunctional component embedded wireless system	e-Textile-based multifunctional systems

nanotubes integrated with microneedle drug delivery (McAllister et. al., 2003) on the same fabric using bio-energy recovery for fiber-based batteries could be integrated into the fabric infrastructure. Adjacency and multifunctional materials address multiple system requirements and reduce or eliminate needs for separate communication and processing components.

A. Integrated Technology Examples

A biodegradable microchip uses a blend of two polymers with different rates of hydrolysis to form membranes that release pulses of drugs over time (Richards-Grayson et al., 2003). It has been suggested that the next advance in such systems may be initiation of drug release from an implant in response to a biosensor using a laser, ultrasonic, or radio-frequency pulse as a trigger (West, 2003). If such a triggering device were incorporated into an e-textile material such as one of the body monitoring garments described previously, and coupled to a polymeric conducting CB sensor, then the chain of sensing, processing, and protective responses could be envisioned.

Power for such a device might be supplied by thin-film lithium ion microbatteries (www.itnes.com), perhaps deposited onto fibers. Successful integration of each of these functions and devices presents many problems, but the fact that the state-of-the-art in separate components is developing rapidly suggests that it may well be a worthy goal (see LaVan et al., 2003, for a summary of recent advances in micro- and nanoscale systems for *in vivo* drug delivery, and Zaugg and Wagner, 2003, for development of methods for large-scale manufacturing of biochips using ink-jet printing).

VIII. PERSPECTIVES

A. Potential of Nanotechnology

The development of nanotechnology for biodefense holds much promise. Many laboratory groups have indicated that progress in both sensors and materials is ongoing and will contribute to the preservation of the environment, assistance to agriculture, enhancements in human health care, and security issues in the near future. This progress is not taking place in a vacuum. Many biodefense technologies will enable, and in turn be enabled by, nanotechnologies. Microelectronics, MEMS, biotechnology, and traditional developments in chemistry will all come into play in this new field. Extensive use is being made of micro- and nanostructures to support or act as detectors. The unique properties of conducting polymers, thin films, nanoparticles, nanoscale surface features, DNAs, and protein structures are being exploited. Self-assembly offers an attractive route for some nanoscale components, although it will not solve all manufacturing problems given the variety of components and enabling requirements.

B. Limitations and Challenges

While the application of nanoscale components to biodefense holds much promise, challenges and limitations must be faced. Practical application of nanosensors and biosensors to widely distributed, inexpensive detector devices will require that laboratory designs be adapted to mass production techniques. Systems will still need to address power requirements (even though they are small) and communication issues, especially when signals must be sent across macroscale distances. Also, current microscale systems such as MEMS and microfluidics chip systems are just now maturing; nanoscale systems are likely to be a decade behind this maturity curve.

In terms of technical challenges, nanoscale systems do not resolve difficult challenges in sampling large volumes of gases or liquids. Also, specific binding sites can be created for many molecules of concern, but the use of predesigned binding sites will not detect unexpected hazardous genetically modified molecules or organisms. The need to differentiate similar malignant and benign organisms such as related strains of bacteria accentuates the problems of specificity and generalized detectability.

Although the bases of many biodefense-relevant sensors are biological binding or reactivity properties — such as the binding of antibodies to the bacterial and viral proteins they recognize or the inhibition of an enzyme by a nerve agent — sensors based on natural macromolecules have certain disadvantages. The use of biological molecules can introduce constraints in sensor operation. The need to keep biological molecules stable and active can produce limits on the operating conditions for sensors, their ruggedness, or the ease of their usage. As a result, nanoscale strategies to construct more stable structures that can mimic biological properties or recognize molecules not readily delivered by natural molecules can make important contributions to sensor improvement.

C. Conclusions

We do not wish to indulge in speculation on how far nanotechnology may push biodefense in the far future, but near-term predictions can be made. The integration of individual components will lead to relatively complicated materials and equipment architectures based on nanotechnology emerging from experiments currently in the laboratory. This will include functional clothing that may support the activities of rescue workers in a disaster, those who must remediate toxic sites, soldiers, police, and the public. Cheap, multifunctional and often tiny detectors will be distributed to monitor a wide variety of parameters from samples to be checked for environmental contamination to the status of individual health. Many examples of excellent laboratory research were highlighted and reviewed in this chapter. The large number of successful results indicates that many routes can and should be supported and explored in the near term for the development of nanotechnological biodefense measures.

REFERENCES

American Vacuum Society, Nanotechnology Innovation for Chemical, Biological, Radiological, and Explosive (CBRE) Detection and Protection, Final Report of Workshop held in Monterey, CA on May 2–3, 2002. http://www.wtec.org/nanoreports/cbre/ (last accessed 12/3/03).

Arnold, B.R., A.C. Euler, A.L. Jenkins, O.M. Uy, and G.M. Murray, Progress in the development of molecularly imprinted polymer sensors, *Johns Hopkins APL Tech. Dig.*, 20, 190–198, 1999.

Baller, M.K. et. al., A cantilever array-based artificial nose, *Ultramicroscopy*, 82, 1–9, 2000.

Barisci, J.N., G.G. Wallace, M.K. Andrews, A.C. Partridge, and P.D. Harris, Conducting polymer sensors for monitoring aromatic hydrocarbons using an electronic nose, *Sensors Chem. Actuators* B84, 252–257, 2002.

Baughman, R.H., A.A. Zakhidov, and W.A. de Heer, Carbon nanotubes: the route toward applications, *Science*, 297, 787–792, 2002.

Belkin, S., Microbial whole-cell sensing systems of environmental pollutants, *Curr. Opin. Microbiol.*, 6, 206–212, 2003.

Besteman, K. et al., Enzyme-coated carbon nanotubes as single-molecule biosensors, *Nano Lett*, 3, 727–730, 2003.

Bonderover, Eitan, Sigurd Wagner, and Zhigang Suo, An inverter woven from flat component fibers for e-textile applications, *Mater. Res. Soc. Symp. Proc.*, 736, H9.10.1–H9.10.7, 2003.

Breadmore, M.C., K.A. Wolfe, I.G. Arcibal, W.K. Leung, D. Dickson, B.C. Giordano, M.E. Power, J.P. Ferrance, S.H. Feldman, P.M. Norris, and J.P. Landers, Microchip-based purification of DNA from biological samples, *Anal. Chem.*, 75, 1880–1886, 2003.

Cadogan, D.P. and L.S. Shook, Manufacturing and performance assessments of several applications of electrotextiles and large-area flexible circuits, *Mater. Res. Soc. Symp. Proc.*, 736, D1.5.1–D1.5.12, 2003.

Cao, Y.C., R. Jin, and C.A. Mirkin, Nanoparticles with raman spectroscopic fingerprints for DNA and RNA detection, *Science*, 297, 1536–1540, 2002.

Cepak, V. et al., Chemical strategies for template syntheses of composite micro- and nano-structures, *Chem. Mater.*, 9, 1065–1067, 1997.

Chakrabarti, R. and A.M. Klibanov, Nanocrystals modified with peptide nucleic acids (PNAs) for selective self-assembly and DNA detection, *J. Am. Chem. Soc.*, 125, 12531–12540, 2003.

Dalton, A.B. et al., Super-tough carbon-nanotube fibers, *Nature*, 423, 703, 2003.

Das, K., J. Penelle, V.M. Rotello, and K. Nusslein, Specific recognition of bacteria by surface-templated polymer films, *Langmuir*, 19, 6226–6229, 2003.

Demers, L.M., D.S. Ginger, S.-J. Park, Z. Li, S.-W. Chung, and C.A. Mirkin, Direct patterning of oligonucleotides on metals and insulators by dip-pen nanolithography, *Science*, 296, 1836–1838, 2002.

Doranz, B., Lipoparticles for drug discovery and diagnostic applications in biodefense, Research, Technologies, and Applications in Biodefense Conference, Washington D.C., August 27–28, 2003, *http://www.healthtech.com/2003/btr/index.htm* (last accessed 12/3/03).

Dubertret, B., M. Clame, and A. Libchaber, Single-mismatch detection using gold-quenched fluorescent oligonucleotides, *Nat. Biotechnol.*, 19, 365–379, 2001.

Duffy, D.J., K. Das, S.L. Hsu, J. Penelle, V.M. Rotello, and H.D. Stidham, Binding efficiency and transport properties of molecularly imprinted polymer thin films, *J. Am. Chem. Soc.*, 124, 8290–8296, 2002.

Dutta, P. et al., Enantioselective sensors based on antibody-mediated nanomechanics, *Anal. Chem.*, 75, 2342–2348, 2003.

Freund, M.S. and Lewis, N.S., A chemically diverse, conducting polymer-based "electronic nose," *Proc. Natl. Acad. Sci. USA*, 92, 2652–2656, 1995.

Guadarrama, A. et. al., Discrimination of wine aroma using an array of conducting polymer sensors in conjunction with solid-phase micro-extraction SPME technique, *Sensors Chem. Actuators*, B 77, 401–408, 2001.

Guan, S., Frequency encoding of resonant mass sensors for chemical vapor detection, *Anal. Chem.*, 75, 4551–4557, 2003.

Gerion, D. et al., Sorting fluorescent nanocrystals with DNA, *J. Am. Chem. Soc.*, 124, 7070–7074, 2002.

Glynou, K., P.C. Ioannou, T.K. Christopoulos, and V. Syriopoulou, Oligonucleotide-functionalized gold nanoparticles as probes in a dry-reagent strip biosensor for DNA analysis by hybridization, *Anal. Chem.*, 75, 4155–4160, 2003.

Haes, A.J. and R.P. Van Duyne, A nanoscale optical biosensor: sensitivity and selectivity of an approach based on the localized surface plasmon resonance spectroscopy of triangular silver nanoparticles, *J. Am. Chem. Soc.*, 124, 10596–10604, 2002.

Hansen, K.M. et al., Cantilever-based optical deflection assay for discrimination of DNA single-nucleotide mismatches, *Anal. Chem.*, 73, 1567–1571, 2001.

Haupt, K., Molecularly imprinted polymers: the next generation, *Anal. Chem.*, 75(17) 377a–383a, 2003.

Hoffman, T.G., G. Canziani, L. Jia, J. Rucker, and R.W. Doms, A biosensor assay for studying ligand–membrane receptor interactions: binding of antibodies and HIV-1 env to chemokine receptors, *Proc. Natl. Acad. Sci. USA*, 97, 11215–11220, 2000.

Huber, D.L., R.P. Manginell, M.A. Samara, B. Kim, and B.C. Bunker, Programmed adsorption and release of proteins in a microfluidic device, *Science*, 301, 352–354, 2003.

Hyun, J., S.J. Ahn, W.K. Lee, A. Chilkoti, and S. Zauscher, Molecular recognition-mediated fabrication of protein nanostructures by dip-pen lithography, *Nano Lett.*, 2, 1203–1207, 2002.

Innovative Biosensors Inc., CANARY™ Technology, http://www.innovativebiosensors.com/tech.htm (11/17/93).

Janata, J. and M. Josowicz, Conducting polymers in electronic chemical sensors, *Nat. Mater.*, 2, 19–24, 2003.

Kasianowicz, J.J., S.E. Henrickson, H.H. Weetall, and B. Robertson, Simultaneous multianalyte detection with a nanometer-scale pore, *Anal. Chem.*, 73, 2268–2272, 2001.

Kirkham, P.M., D. Neri and G. Winter, Toward the design of an antibody that recognises a given protein epitope, *J. Mol. Biol.*, 285, 909–915, 1999.

Kong, J. et al., Nanotube molecular wires as chemical sensors, *Science*, 287, 622–625, 2000.

Koper, O.B. et al., Nanoscale powders and formulations with biocidal activity toward spores and vegetative cells of Bacillus species, viruses, and toxins, *Curr. Microbiol.*, 44, 49–55, 2002.

Kumar, A., Biosensors based on piezoelectric crystal detectors: theory and application, *JOM-e*, 52, 10, 2000.

LaVan, D., T. McGuire, and R. Langer, Small-scale systems for in vivo drug delivery, *Nat. Biotechnol.*, 2110, 1184–1191, 2003.

Lee, K.-B., S.-J. Park, C.A. Mirkin, J.C. Smith, and M. Mrksich, Protein nanoarrays generated by dip-pen nanolithography, *Science*, 295, 1702–1705, 2002.

Levitan, B., Stochastic modeling and optimization of phage display, *J. Mol. Biol.*, 277, 893–916, 1998.

Lim, D.V., Detection of microorganisms and toxins with evanescent wave fiber-optic biosensors, *Proc. IEEE*, 91, 902–907, 2003.

Lin, D.Y. et al., Tailored nanofiber morphologies using modulated electrospinning for biomedical applications, *Mater. Res. Soc. Symp. Proc.*, 736, D3.8.1–D3.8.6, 2003.

Liu, J. and Wan, M., Synthesis, characterization and electrical properties of microtubules of polypyrrole synthesized by a template-free method, *J. Mater. Chem.*, 11, 404–407, 2001.

MacDiarmid, A.G., W.E. Jones, Jr., I.D. Norris, J. Gao, A.T. Johnson, Jr., N.J. Pinto, J. Hone, B. Han, F.K. Ko, H. Okuzaki, and M. Liguno, Electrostatically-generated nanofibers of electronic polymers, *Synth. Met.*, 119, 27–30, 2001.

Marculescu, D., R. Marculescu, S. Park, and S. Jayaraman, Ready to ware, *IEEE Spectrum*, 40(10) 28–32, October 2003.

Martin, C.R., Nanomaterials: a membrane-based synthetic approach, *Science*, 266, 1961–1966, 1994.

Mazzola, L., Commercializing nanotechnology, *Nat. Biotechnol.*, 2110, 1137–1143, 2003.

McAllister, D.V. et al., Microfabricated needles for transdermal delivery of macromolecules and nanoparticles: fabrication methods and transport studies, *Proc. Natl. Acad. Sci. USA*, 10024, 13755–13760, 2003.

Moulin, A., S. O'Shea, and M. Welland, Microcantilever-based biosensors, *Ultramicroscopy*, 82, 23–31, 2000.

Natarajan, K. et al., Electrotextiles: present and future, *Mater. Res. Soc. Symp. Proc.*, 736, D3.10.1–D3.10.6, 2003.

Novak, J.P., E.S. Snow, E.J. Houser, D. Park, J.L. Stepnowski, and R.A. McGill, Nerve agent detection using networks of single-walled carbon nanotubes, *Appl. Phys. Lett.*, 83, 4026–4028, 2003.

Panchapakesan, B. et al., Nanoparticle engineering and control of tin oxide microstructures for chemical microsensor applications, *Nanotechnology*, 12, 336–349, 2001.

Park, S.J., A. Taton, and C.A. Mirkin, Array-based electrical detection of DNA with nanoparticle probes, *Science*, 295, 1503–1506, 2002.

Park, S. and S. Jayaraman, Smart textiles: wearable electronic systems, *MRS Bull.*, 288, 585–591, 2003.

Parthasarathy, R.V. and C.R. Martin, Template-synthesized polyaniline microtubules, *Chem. Mater.*, 6, 1627–1632, 1994.

Pavlou, A., A.P.F. Turner, and N. Magan, Recognition of anaerobic bacterial isolates *in vitro* using electronic nose technology, *Lett. Appl. Microbiol.*, 35, 366–369, 2002.

Petrenko, V.A. and V.J. Vodyanoy, Phage display for detection of biological threat agents, *J. Microbiol. Meth.*, 53, 253–262, 2003.

Post, E.R., M. Orth, P.R. Russo, and N. Gershenfeld, E-broidery: design and fabrication of textile-based computing, *IBM Syst. J.*, 39(3&4) 840–860, 2000. http://www.research.ibm.com/ journal/sj/393/part3/post.html (accessed 11/25/03).

Qiu, H., M. Wan, B. Matthews, and L. Dai, Conducting polyaniline nanotubes by template-free polymerization, *Macromolecules*, 34, 675–677, 2001.

Richards-Grayson, A.C. et al., Multi-pulse drug delivery from a resorbable polymeric microchip device, *Nat. Mater.*, 211, 767–772, 2003.

Rider, T.H. et al., B cell-based sensor for rapid identification of pathogens, *Science*, 301, 213–215, 2003.

Sadik, O.A., Bioaffinity sensors based on conducting polymers: a short review, *Electroanalysis*, 1112, 839–844, 1999.

Sasaki, D.Y. et al., Crown ether functionalized lipid membranes: lead ion recognition and molecular reorganization, *Langmuir*, 18, 3714–3721.

Sblattero, D. and A. Bradbury, Exploiting recombination in single bacteria to make large phage antibody libraries, *Nat. Biotechnol.*, 18, 75–80, 2000.

Schreuder-Gibson, H.L. et al., Chemical and biological protection and detection in fabrics for protective clothing, *MRS Bull.*, 288, 574–578, 2003.

Schreuder-Gibson, H.L. et al., Protective textile materials based on electrospun nanofibers, *J. Adv. Mater.* 343, 44–55, 2002.

Schreuder-Gibson, H.L., J.E. Walker, W. Yeomans, and R. Stote, Development and applications of electrospun membranes, *Proceedings of Membranes Conference*, Newton, MA, December 2–4, 2003.

Semancik, S. et al., Micro-hotplate platforms for chemical sensor research, *Sensors Actuators B*, 771, 579–591, 2001.

Sepaniak, M.J., P. Datskos, N. Lavrik, and C.A. Tipple, Microcantilever transducers: a new approach in sensor technology, *Anal. Chem.*, 74(21) 568A–575A, 2002.

Shah, R.R. and N.L. Abbott, Principles for measurement of chemical exposure based on recognition-driven anchoring transitions in liquid crystals, *Science*, 293, 1296–1299, 2001.

Shi, H., W.-B. Tsai, M.D. Garrison, S. Ferrari, and B.D. Ratner, Template-imprinted nanostructured surfaces for protein recognition, *Nature*, 398, 593–597, 1999.

Shipway, A.N., E. Katz, and I. Willner, Nanoparticle arrays on surfaces for electronic, optical, and sensor applications, *Chem. Phys. Chem.*, 1, 18–52, 2000.

Star, A., J.-C. P. Gabriel, K. Bradley, and G. Gruner, Electronic detection of specific protein binding using nanotube FET devices, *Nano Lett.*, 3, 459–463, 2003.

Stratis-Cullum, D.N., G.D. Griffin, J. Mobley, A.A. Vass, and T. Vo-Dinh, A miniature biochip system for detection of aerosolized *Bacillus globigii* spores, *Anal. Chem.*, 75, 275–280, 2003.

Su, M., S. Li, and V.P. Dravid, Miniaturized chemical multiplexed sensor array, *J. Am. Chem. Soc.*, 125, 9930–9931, 2003.

Tao, A., F. Kim, C. Hess, J. Goldberger, R. He, Y. Sun, Y. Xia, and P. Yang, Langmuir–Blodgett silver nanowire monolayers for molecular sensing using surface-enhanced Raman spectroscopy, *Nano Lett.*, 1229–1233, 2003.

Thundat, T. and Brown, G.M., Environmental monitoring using cantilever sensors, *Oak Ridge Natl. Lab. Life Sci. Technol.*, 58–62, 2002.

Tipple, C. A. et al., Nanostructured microcantilevers with functionalized cyclodextrin receptor phases: self-assembled monolayers and vapor deposited films, *Anal. Chem.*, 74, 3118–3226, 2002.

Wagner, G.W., P.W. Bartram, O. Koper, and K.J. Klabunde, Reactions of VX, GD, and HD with nanosize MgO, *J. Phys. Chem.*, B103, 3225–3228, 1999.

Wang, J. et al., Single-channel microchip for fast screening and detailed identification of nitroaromatic explosives or organophosphate nerve agents, *Anal. Chem.*, 74, 1187–1191, 2002a.

Wang, J. et al., Measurements of chemical warfare agent degradation products using an electrophoresis microchip with contactless conductivity detector, *Anal. Chem.*, 74, 6121–6125, 2002b.

Wang, X., C. Drew, S.-H. Lee, K.J. Senecal, J. Kumar, and L.A. Samuelson, Electrospun nanofibrous membranes for highly sensitive optical sensors, *Nano Lett.*, 2, 1273–1275, 2002.

West, J.L., Drug delivery: pulsed polymers, *Nat. Mater.*, 211, 709–710, 2003.

Williams, K.A. et al., Carbon nanotubes with DNA recognition, *Nature*, 420, 761, 2002.

Wu, G. et al., Bioassay of prostate-specific antigen PSA using microcantilevers, *Nat. Biotechnol.*, 19, 856–860, 2001.

Xiao, Y. et al., Plugging into enzymes: nanowiring of redox enzymes by a gold nanoparticle, *Science*, 75(5) 299, 1877–1881, 2003.

Yinon, J., Detection of explosives by electronic noses, *Anal. Chem.*, 99A–105A, 2003.

Yoon, H.J. et al., Solid-state ion sensors with a liquid junction-free polymer membrane-based reference electrode for blood analysis, *Sensors Actuators B*, 64, 8–14, 2000.

Zaugg, F.G. and P. Wagner, Drop-on-demand printing of protein biochip arrays, *MRS Bull.*, 2811, 837–842, 2003.

Zimmerman, S.C., M.S. Wendland, N.A. Rakow, I. Zharov, and K.S. Suslick, Synthetic hosts by monomolecular imprinting inside dendrimers, *Nature*, 418, 399–403, 2002.

CHAPTER 6

Social and Economic Contexts: Making Choices in the Development of Biomedical Nanotechnology

Ineke Malsch

CONTENTS

I. INTRODUCTION

In this chapter on social and economic aspects related to the emergence of biomedical nanotechnology, I take a different angle from the rest of this book. I start from the perspective of our global society and the needs for better and affordable health care of ordinary people in different parts of our world. From there, I zoom into the priorities of present-day nanotechnology research for biomedical applications. I hope the combination of these perspectives will lead to constructive dialogues among nanotechnology researchers, other promoters of the science, and the general public that will contribute to more efficient development of biomedical applications of nanotechnology that can solve real needs of real people. I also sketch the health care and technology development systems that form the context for present development and eventual use of the biomedical nanotechnologies described earlier in this book. Because I am a European, my analysis will be most relevant to the European context, but I include information and discussions about the United States (U.S.) and other countries.

The starting point of my analysis is the most pressing need for health care worldwide. Therefore, it makes sense to call to mind the United Nations (UN) Millennium Development Goals that form a global framework of actions aiming to fill this need. In 2000, the General Assembly of the United Nations adopted its Millennium Declaration; the countries represented promised to work together to establish a more peaceful, prosperous, and just world. Among other issues, they set eight Millennium Development Goals intended to be reached by 2015.

One goal is to halt and begin to reverse the spreads of HIV/AIDS, malaria, and other major diseases that afflict humanity. Another goal is to develop a global partnership for development involving not only governments, but also the private sector and civil society. In particular, the UN General Assembly wants:

> To encourage the pharmaceutical industry to make essential drugs more widely available and affordable by all who need them in developing countries; to develop strong partnerships with the private sector and with civil organizations in pursuit of development and poverty eradication; [and] to ensure that the benefits of new technologies, especially information and communication technologies ... are available to all.[1]

In this chapter, I look into presently expected societal and economic benefits of biomedical nanotechnology and how these priorities relate to these Millennium Development Goals or can be adapted to them.

II. GLOBAL TRENDS IN HEALTH CARE NEEDS

In 2002, the World Health Organization (WHO) published its *World Health Report* dedicated to the 25 key global health risks. This report shows a big difference in health and healthy life expectancy for people in the northern and southern hemispheres. The top ten health risks[2] are:

1. Underweight
2. Unsafe sex
3. High blood pressure
4. Tobacco consumption
5. Alcohol consumption
6. Unsafe water, sanitation, and hygiene
7. Iron deficiency
8. Indoor smoke from solid fuels
9. High cholesterol
10. Obesity

A. Social and Economic Damages Arising from Disease

The WHO recommends that governments develop risk prevention policies mainly focused at educating people to change unhealthy behaviors such as poor eating habits, unsafe sex, smoking, and alcohol use. WHO also proposes cost-effective treatments involving existing drugs. The WHO believes that its prevention strategies may lead to increases of 5 to 10 years in healthy life expectancy for people in the developed world and in developing countries, respectively. The WHO report does not deal with the issues of developing new drugs or medical technologies that will not be affordable for people in developing countries for several decades because of current industrial property rights legislation. Why then do we need biomedical nanotechnology? Principally because many people only start worrying about their health after they become ill, when prevention is of little help. Furthermore, people are always vulnerable to infectious diseases and accidents and can suffer health effects of genetic disorders. Nanotechnology also contributes to the development of prostheses and implants that give disabled people a better quality of life.

B. Diseases

1. Infectious Diseases

Looking at the state of world health from a different angle, infectious diseases are very real threats to the lives and health of people in all parts of the world. Table 6.1 lists most common diseases and numbers of victims. HIV/AIDS alone represents the fourth major cause of death — 2.9 million deaths in 2000.[2] HIV is a retrovirus or type of ribonucleic acid (RNA) virus. RNA is responsible for protein expression inside cells. The infection is spread through direct contact of bodily fluids. Unsafe sex and blood transfusions are the main causes of infections. At present, medication that can control the disease is available, and this enables patients to lead healthy lives for many years longer than without the medication. Unfortunately, the drugs are very expensive and not available to most patients in developing countries. No anti-AIDS vaccine or drug that will cure HIV/AIDS completely is yet available.

In developing countries, HIV/AIDS, malaria, tuberculosis, and other tropical diseases claim many victims and lead to considerable losses of national income. Malaria alone kills 1 million people each year and infects many more.[3] Malaria is caused by protozoa — animal parasites — and transmitted by infected mosquitoes. The disease has been eradicated in western countries by elimination of the malaria mosquitoes but it is still endemic in developing countries in tropical areas. The available medication consists of strong drugs prescribed to travelers. Local populations can protect themselves by using insecticide-treated nets. The WHO and other organizations are attempting to stimulate research to develop better antimalarial drugs.

In western countries, diseases such as influenza, legionnaires' disease, and antibiotic-resistant infections claim many victims every year, especially among sick and elderly people. Additional risks are the emergence of new diseases and the ability of relatively innocent existing diseases to evolve into more deadly variants. This frequently happens with animal diseases that affect humans (zoonoses). Continuous risks of zoonoses are present in areas of intensive livestock farming. Bovine spongiform encephalitis (BSE or mad cow disease) is an example of a zoonotic disease that can cause variant Creutzfeldt–Jakob disease in humans — a lethal disease that causes spongy brain damage. Between 1996 and 2002, 139 cases were reported worldwide and no cure exists to date.[4]

Table 6.1   Diseases and Numbers of Victims

Disease	Annual Mortality	Number Infected	Year	Source
Cardiovascular diseases	Approximately 17 million		2003	WHO
Cancer	6.2 million		2000	WHO
HIV/AIDS	3 million	38 million	2003	UN, July 2004
Malaria	1 million	300 million acute cases per year	2003	WHO
Tuberculosis	2 million	9 million per year	2002	WHO

SARS is an infection caused by a coronavirus, a type of RNA virus. The virus can be spread by droplets and causes infection of the lungs. Experts believe SARS is a recombinant animal virus that has changed itself so that it is now infectious and even deadly to humans. The severe acute respiratory syndrome (SARS) epidemic in 2003 and the threats of bioterrorism and biological warfare clearly show the continuing needs for new or improved antibiotics, vaccines, and rapid diagnostic tests for identifying dangerous viruses and bacteria.

Bioterrorists and even certain countries are believed to be able and willing to use biological weapons of mass destruction such as anthrax, smallpox, botulinum toxin, and Ebola virus. The U.S. and more recently governments of other countries[5] are funding biodefense research to develop sensors to identify and vaccines to defend against such biological weapons. They are also developing sensors for detecting nuclear and chemical agents.

a. Types of Infectious Diseases and Treatments

We must distinguish viral and bacterial diseases. Bacteria are living microorganisms. They are complete cells that can replicate as long as they have suitable and sufficient food supplies. The two types of drugs able to fight bacterial infections are specific antibiotics and broad antibiotics that are effective against several different infections. The need for new antibacterial drugs continues because bacteria tend to become immune to antibiotics. A virus consists of a strand of DNA or RNA that requires a host cell to be able to replicate. Antiviral drugs developed to date are protease inhibitors that reduce the activities of the enzymes that replicate the virus strands. A protease inhibitor can be developed only in the presence of a specific viral protease — development of a protease takes about 10 years before the product can enter the market. Therefore it is not possible to quickly develop antiviral drugs against unknown emerging diseases such as SARS. The need for broad antiviral drugs that are effective against multiple viruses is especially pressing, based on an interview with Willy Spaan of *Chemical 2 Weekly*, a Dutch magazine for Chemists in the April 2003 issue. As noted earlier, nanotechnology can contribute to filling these needs by incorporation of advanced genomics, proteomics, and drug discovery techniques in laboratory instruments and developing better and more economical diagnostic methods.

2. Cancer

In 2000, 10 million people developed new cancers and 6.2 million people died of the disease. The WHO fears that by 2020, 15 million people will develop new cancers annually. However, preventive actions related to smoking, diet, and control of infections can prevent a third of new cases. Another third may be curable by then. The U.S., Italy, Australia, Germany, The Netherlands, Canada, and France had the highest overall cancer rates in 2000. Technical solutions include "early detection through screening, using methods such as mammography, magnetic resonance, or computed tomography .... Molecular genome research will reveal a tremendous amount of information, but it is not clear how easily these discoveries will translate into actual lives saved and may well be restricted to rare cancers.... The medical community must develop a wide spectrum of tests for other cancers (than cervical cancer) and are now evaluating many procedures to determine if they are effective and practical."[6]

A major pull factor for nanotechnology research is the need for new or improved anticancer drugs. Governments and private foundations invest large amounts of money in anticancer research. Nanoscale drug delivery is especially useful for anticancer treatments. Most of the available chemotherapy and radiation therapies are toxic both to cancer cells and to the rest of the human body. More effective cures with fewer side effects require targeted delivery and controlled release of the medication within the cancer. Several types of nanodrug delivery systems are promising, as discussed in Chapter 2, this volume.

Diagnostics that include nanotechnology are also important for cancer patients because economical and easy-to-use biochips that can test for different kinds of cancers will allow the disease to be discovered earlier and this will increase the chances for successful treatment. Also, imaging techniques involving nanoparticles that can detect and target cancer cells or tissues outside the body appear promising.

3. Cardiovascular Diseases

Cardiovascular diseases cause approximately 17 million deaths annually. The WHO estimates that more than half of these deaths can be prevented by healthier lifestyles and better quality processed foods that contain less salt.[2] In Western countries, heart patients can benefit from treatments such as stents, pacemakers, and even heart transplants. Nanotechnology can contribute to safer stents by implementing biocompatible surface layers or by including slow release drugs to combat rejection. Pacemakers already include nanostructured materials in the electrodes that deliver electrical shocks to the heart. At the moment, transplants can only be accomplished with human hearts.

Xenotransplantation is implantation of one or more cells or even an entire organ from one species into another. For example, xenotransplantation of the hearts of specially bred pigs into humans is one future option, but not an unproblematic one. The two risks of xenotransplantation are rejection of the transplanted cells or organ by the host and the risk of zoonoses — infection of humans by mutations of animal-specific infectious diseases. The SARS epidemic shows how real this risk is because SARS is probably caused by a mutated animal-specific coronavirus. In the long term, nanotechnology may contribute to the development of artificial tissues or even whole hearts and other organs that can be produced under sterile conditions and include surface layers that are compatible with each patient's immune system.

4. Other Diseases

Similar xenotransplantation approaches are relevant for liver and kidney replacements. Nanotechnology can play a role in artificial organs implanted within the body and be utilized for membranes and other components or materials incorporated in external dialysis instruments.

The brain is vulnerable to genetic disorders such as Parkinson's, Alzheimer's, and Huntington's disease and also to brain damage caused by accidents or cancers. As the average life expectancy in Western countries increases, more people suffer from chronic age-related diseases that are not lethal but cause losses of quality of life. Parkinson's and Alzheimer's are among the better known diseases that emerge as people live longer. Nanotechnology can be used to develop drug delivery mechanisms across the blood–brain barrier and for gene therapy; single animal cells can be transplanted into a patient to enhance the production of serotonin against Parkinson's disease. One way to avoid the risk of zoonoses from cell xenotransplantation is microencapsulation of a single cell or a cluster of cells in an artificial shell. In this way, the animal cells are not in direct contact with human tissue but can still perform the functions for which they were implanted in the body. This technique may be suitable for transplanting single cells but is not so relevant to implantations of entire organs.

Orphan diseases are chronic or lethal disorders affecting only a small part of the population — fewer than 1 person in 2000.[7] These diseases are often genetically determined. Experimental drugs and therapies are under development in research laboratories and start-up companies because the drugs or treatments will be relatively expensive and the developers must find a niche where their products will not have to compete with existing drugs. Research is funded by special government or private funds. For several orphan diseases, gene therapy is a potential cure. Nanotechnology can contribute by developing drug delivery vectors or by contributing to diagnostic lab-on-a-chip techniques and high throughput screening for drug discovery.

C. Disabilities

Many people suffer disabilities arising from birth, determined genetically, or resulting from accidents. Prostheses and implants can help these patients lead lives that are as normal as possible. Medical technology that is integrated into the human body is not the only solution available. Patients can also use other technologies and skills, such as wheelchairs for those with ambulation problems, Braille for the blind, and lip reading and sign language for the deaf people as alternatives.

1. Blindness and Visual Impairments

Worldwide, there are 180 million visually impaired people including about 40 to 45 million blind people. The WHO estimates that nine out of ten blind people live in developing countries. Blindness can be attributed to cataracts (clouding of the lenses, 46%), trachomas (eyelid infections, 12.5%), childhood onset (3.3%), onchocerciasis (river blindness, 0.6%), and other causes. Based on a number of factors including the aging of populations, the WHO expects the number of blind people to total 100 million worldwide by 2020.

To reverse this trend, WHO implemented Vision 2020, a global program that aims to eliminate avoidable blindness (about 80% of the total) by 2020. This program is more concerned with building health care facilities to treat patients in developing countries and with dissemination of existing technologies than with futuristic devices such as electronic eyes. One aim is to use intraocular lenses to cure cataracts by replacing natural lenses with artificial ones. The first priorities of the program are prevention and treatment of cataracts, trachomas, onchocerciasis, childhood blindness, and refractive errors and low vision.[8–11] Nanotechnology may not have an obvious role to play in solving these problems.

a.   Artificial Eyes

Attempts to develop electronic eyes or retina implants to restore sight are ongoing. The first experimental electronic eye was placed in a blind person in 1978 by the Dobelle institute. In 2002, the institute implanted eight more patients with improved versions of the electronic eye. The technology is still crude and includes sensors in glasses, implanted platinum electrodes, and a laptop computer to process the signal.[12] The technology is experimental and is available only to patients who can afford it. Some safety issues have also been raised. Another approach to curing blindness involves nanotechnology.

Several ongoing projects in the U.S. and Germany involve academic groups and small and medium-sized enterprises working on retina implants and several small companies that have developed retinal implants and are now performing clinical trials of their systems. The current technologies are still microsystems but they include nanotechnology in specific elements such as electrodes.[13] The first commercial implants are expected on the market by 2008.

b.   Paralysis and Prostheses

Paralysis is caused by accidents or other injuries to the spinal cord and central nervous system. Clearly, nanotechnology can find applications in this area, especially related to tissue engineering and neural cell growth stimulation. Many people already have external or implanted prostheses. Hip replacements in elderly people are the most common forms. These devices sometimes cause rejection or other problems because they fit poorly. Nanostructured surfaces can help to improve the growing in of artificial bone within the body (see Chapter 3, this volume).

D. Discussion

The most common life-threatening conditions worldwide are cardiovascular diseases, cancer, and infectious diseases. Nanostructured materials may be included in the future in medical devices such as pacemakers and drug eluting stents to treat cardiovascular diseases. Currently, no cures exist for certain forms of cancer and the chemotherapy and radiation used to treat other types of cancer produce severe side effects. Nanodrug delivery may help to reduce such side effects. Nanoparticles applied in new imaging techniques and diagnostic chips may help identify cancers and other diseases in early phases of development so they may be easier to cure. In general, nanotechnology may contribute to faster development of new active drug compounds by applying lab-on-a-chip techniques to high throughput screening in the pharmaceutical industry. Many people suffer from handicaps and nonlife-threatening diseases. Applications of biomedical nanotechnology in prostheses and active and passive implants may allow these patients a better quality of life.

III. HEALTH CARE SYSTEMS: TRENDS AND ECONOMICS

This section will sketch general trends in health care systems around the world as background to developments in biomedical nanotechnology. The populations of Western countries and more advanced developing countries are aging because of higher per capita income and better quality of life. Older people tend to suffer more age-related diseases, as a result of which they become more fragile and need more care. This leads to increased costs of health care systems. Another cost-increasing trend arises from the success of pharmaceutical and medical technology developments. In particular, the rapid progress in biotechnology, genomics, and proteomics-based drug compound screening and development is leading to the availability of treatments for formerly untreatable diseases. This increases the direct cost of health care because medications must be paid for and are expensive during their 20 years on the market while they are still protected by patents. The emerging questions are whether we are willing to pay for all that is technically possible and, if not, what are our priorities for 21st century health care systems?

A. Health Care Market

Health care includes pharmaceuticals and medical technologies. The main actors in the health care market are governments, public and private health care insurance companies, suppliers of pharmaceuticals and medical technologies, medical professionals, patients and consumers, and outsiders (Table 6.2). Governments are responsible for organizing national health care systems; for financing the infrastructure and care; and for regulation. The governments and the health care insurance companies decide which care they will reimburse. The suppliers determine new drugs and technologies to develop and produce. Medical professionals, especially doctors, decide which drugs or technologies to prescribe to patients; patients and consumers are more decisive because they are better informed about alternative medications. Outsiders are uninsured people in developed countries and people in least developed countries who have no access to health care markets.

At this early stage of nanotechnology development, the market for health care is not directly relevant. However, if such development is to be demand-driven, nanotechnology researchers should take into account the general trends in this market. This means that cost–benefit analyses and the views of stakeholders must be considered at early stages of decision making that determines R&D priorities. Ascertaining which technological developments can potentially deliver the most benefits to world health for the least investment is the key question that should guide the decisions of policy makers.

Table 6.2 Health Care Market Stakeholders and Their Roles

Governments	Insurance Companies	Medical and Pharmaceutical Industries	Medical Professionals	Patients and Consumers	Outsiders
Organize national health care systems	Insure	Manufacture drugs and medical devices	Prescribe drugs and medical devices	Receive care and medications	People in developing countries or uninsured individuals
Finance infrastructures and care	Determine which services, medications, and technologies to refund	Develop new drugs and medical devices	Use medical technologies	Co-decide on prescriptions and technologies	
Regulate				Engage in self-medication	
Insure					
Finance R&D efforts					

B. Characteristics of Health Care Systems

The health care insurance system in the U.S. is market dominated; the public sector is more dominant in Europe. All citizens in Japan have compulsory health insurance. The U.S. system is highly innovative, and is usually the first in the world to incorporate new drugs, medical technologies, and practices. It is also by far the most expensive system based on percentage of gross domestic product (GDP; 14% in 2001 compared to 8% average in the Organization for Economic Cooperation and Development). At the same time, 14% of the U.S. population is uninsured.

Current trends include increasing costs of prescription drugs, due primarily to the market introductions of new drugs for illnesses for which no medications existed earlier and to a switch from other forms of treatment to prescribing drugs. Experts in health care economics disagree about the cost effects of introducing new medical technologies. The prices of new technologies are higher for early adopters such as the U.S. Cutler and others[14,15] have found evidence that the cost increases are partly offset by reductions in other health care costs when diseases are avoided.[16] This implies that the innovation environment in the U.S. is more stimulating for start-up SMEs developing biomedical nanotechnology products or incorporating nanocomponents or materials into new drugs or medical devices. The U.S. is also more stimulating for the R&D departments of large pharmaceutical companies, as reflected by the relative abundance of such companies located there. At the same time, regulations for market access (FDA approval) are stricter in the U.S., but this serves to enhance quality in other countries.

In Europe, health care policies are mainly organized on a national government basis, even by members of the European Union (EU). The EU is responsible for harmonization of legislation, primarily by imposing directives and other types of legislation. Directives are subsequently implemented in the national laws of the EU member states. The pharmaceutical and medical devices industries are regulated separately. The cost and organization of the health care sectors have been debated for over a decade in many countries.

Japan has a system of compulsory health insurance for all and enjoys the highest life expectancy worldwide and a relatively inexpensive health care system. The high life expectancy may be related to healthy diets. Trends in new technology and drug development do not play a role in discussions about changes in health care economics in Japan.[17]

Especially in least developed countries, preventable diseases and disabilities cause considerable damage to national economies. The most pressing examples are HIV/AIDS, malaria, tuberculosis, and other tropical diseases that have produced catastrophic impacts on the economies and societies of sub-Saharan Africa. In the poorest countries, the issue is how to escape the poverty trap that prevents their populations from having ample food, clean water, sanitation, and housing to stay healthy enough to earn decent incomes that would enable them to pay for medical care and the basics required for leading healthy lives. These countries lack both national health care systems and provisions for health insurance. In September 2003, the international community reached an agreement in the course of World Trade Organization negotiations that will allow imports of inexpensive generic alternatives to expensive patented drugs in developing countries that do not produce the drugs in question.

C. Discussion

Developed countries offer still the most obvious market opportunities for innovative health care products, particularly because the aging populations of critical and insured health care consumers may lead to more demand for pharmaceuticals and medical devices that utilize nanotechnology. The increased use of nanotechnology and other innovations in health care may be hampered if it leads to rising costs. Politicians and insurance companies are already confronted with difficult choices in health care priorities. Nanotechnology will have to compete with other technical and nontechnical options.

The most pressing global needs for health care exist in countries that lack basic levels of national health care systems. Possible solutions in those countries do not concern new nano or other technologies; they require investments in health care workers, hospitals, local availability of sufficient supplies of essential drugs, and basic sanitation measures. Exceptions in which biomedical nanotechnology may be useful include high throughput screening technologies used to develop drugs to combat major infectious diseases and rapid and economical diagnostics. Water purification or desalination may also benefit from the use of nanomaterials, for example, for ultrafiltration membranes that can ensure safe water supplies and ultimately healthier populations. However, a discussion of these applications of nanotechnology is beyond the scope of this book.

D. Markets for Biomedical Nanotechnology R&D

1. Pharmaceutical R&D

There are two distinct markets for biomedical nanotechnology R&D. The first and largest is the market for pharmaceutical R&D. This market is dominated by large pharmaceutical industries that fund most of the R&D costs of new drugs in house and produce and sell the drugs after they are allowed on the market (Table 6.3). Governments play dual roles. On the one hand they fund fundamental R&D and R&D related to orphan and other diseases that are not attractive targets for pharmaceutical companies.

Governments also regulate market access of new drugs and act as deciders about insurance coverage for such drugs. Private charities and national research funding councils fund research on new drugs and biomedical devices to treat diseases that are their priorities, for example, the cancer research foundations. Universities, specialized medical research centers, and academic hospitals carry out more fundamental research on new drugs. For the past decade or so, a number of SMEs have acted as intermediaries between these institutions and the pharmaceutical industries. They tend to take patented results of academic research further toward new drugs or drug screening technologies. The SMEs and some larger companies develop the products, then produce and sell them or big companies may pick up such technology if it is successful. Patents are vital parts of the markets for pharmaceutical R&D. As a result, patenting organizations, lawyers, and advisors are also important players.

a. Patents

Patents for new active drug compounds play a large role in the market for pharmaceuticals because the R&D investments are high and few drug candidates make it to the market. According to the Pharmaceutical Research and Manufacturers of America, a pharmaceutical industry association, only 1 of every 5000 active compounds tested makes it to the market as a drug.[18] A patent grants an exclusive right to market a product for 20 years. Pharmaceutical companies claim to require this time to recover their R&D costs and cover the risks of failure of a drug candidate. Patenting involves a number of political issues. European patents are much more expensive than U.S. patents and involve much more "red tape" such as the costs of translation into all the national languages of the individual countries in which applications are filed. This puts European companies at a competitive disadvantage and may be one reason why new drugs are marketed earlier in the U.S. Patented drugs are expensive; patients in developing countries and uninsured people cannot acquire the treatments they need because of the expense. A number of initiatives have been proposed to subsidize essential drugs or have companies offer them at lower prices in developing countries,[19] but these initiatives are accompanied by the risks of parallel imports of cheaper versions to be sold for official prices in rich countries and fraud. Policymakers in Western countries are trying to prevent these problems.

b. Global Competition

The pharmaceutical industry plays an important role in the global competition among Europe, the U.S., and Japan. The EU is developing new legislation to foster "a stronger European-based pharmaceutical industry for the benefit of the patient."[20] One of the main elements of EU policies affecting the pharmaceutical sector is "the need to strengthen the competitiveness of the European pharmaceutical industry, with particular regard to encouraging research and development."[21] The U.S. is clearly the world leader in venture capital investment in health care and biotechnology. The amount of investment has varied between two and five times the investment in the entire EU from 1995 to 2000 (see Table 6.3).

Government funding for health R&D shows the same pattern. "Most European governments invest less than 0.1% of GDP in health R&D; this compares to the U.S. figure of 0.19%. In 2000, the U.S. government invested nearly five times more in health R&D than the fourteen EU countries for which figures are available. This is almost $21 billion. The EU budget for life sciences, genomics, and biotechnology for health in the Sixth Research Framework Program amounts to $2,255 million for the period 2003–2006, i.e., $564 million per year on average."[20] U.S.-based pharmaceutical companies overtook European companies in R&D expenditures during the 1990s. In 1990 through 1992, European pharmaceutical companies invested more than their U.S.-based competitors. Between 1993 and 1996, their budgets were more or less the same, but from 1997 until 2000, the annual growth rate of R&D investment of U.S.-based companies was faster than the rates of European companies. Most European companies are headquartered in the United Kingdom, Germany, and France. Japanese companies invested considerably less in R&D.[20] Europe leads only in employment levels. Between 1990 and 2000, the pharmaceutical industry in the EU employed almost 500,000 people (each year), compared to around 200,000 people in the U.S. and Japan.[20] The pharmaceutical industry is the EU's fifth largest industrial sector.

The European Commission intends to foster competitiveness of this important sector by a number of policy measures, including improving access to innovative medicines and a more transparent approach to the assessment of new medicines by improving dialogues during development. The commission intends to fund the development of innovative medicines through the Sixth Framework thematic program on "life sciences, genomics, and biotechnology for health." The commission also intends to strengthen the European science base by stimulating networking in the form of "virtual institutes of health" and by setting up a European Center for Disease Prevention and Control. One area of particular concern is targeting communicable diseases prevalent in developing countries via the European Developing Countries Clinical Trial Partnership (EDCTP). The commission also intends to put into practice its "Life Science and Biotechnology Action Plan" that aims to foster a European biotechnology industry and stimulate dialogues between the public and the life sciences sectors. The European Commission has also developed an action plan focused on science and society.

Table 6.3   Pharmaceutical and Medical Device R&D Market Stakeholders and Their Roles

Intergovernmental Organizations	Governments	Pharmaceutical and Medical Device Manufacturers	Charities and National Research Funding Agencies	Universities and Research Centers	Patenting Organizations	SMEs	Standards Bodies	Insurance Companies	Medical Professionals
WHO monitors world health; coordinates national health care campaigns and health care in least developed countries OECD coordinates national health care R&D and economic policies WTO coordinates regulation of world trade in medical devices EU harmonizes market access and legislation on medical devices in EU member states	Fund R&D and education; determine economic policy; regulate market access; set health care policy; organize national health care systems; conduct foreign affairs; finance infrastructure and care; regulate; insure; finance R&D	Manufacture pharmaceuticals and medical devices; finance and conduct R&D for new product development	Finance R&D; charity agencies lobby for patients' interests	Educate researchers and medical personnel; conduct precompetitive R&D; spin off medical and life science start-ups	Patent new medical devices, drugs, and technologies	Develop new platform technologies and products; conduct contract R&D; network with pharmaceutical industries	Organize debates about needs for new standards; publish standards; conduct measurements and standardization tests	Insure; decide which services, medications, and technologies to refund	Prescribe drugs and medical devices; use medical technologies

Patients and Consumers

Receive medical care and medications; co-decide on prescriptions and technologies to use; engage in self-medication

Outsiders

No roles; people in developing countries and uninsured individuals

E. Orphan Drugs and Tropical Diseases

Large pharmaceutical companies do not invest in orphan drugs or treatment of tropical diseases unless their R&D is heavily subsidized or consumers will reward their socially responsible enterprises. Without subsidies or consumer support, the companies cannot recover their investments or make profits. Innovation to develop new drugs for tropical diseases is funded by United Nations organizations and charities or by the EU and national governments in Western countries. The European Commission's EDCTP 2003–2007 effort aims to develop affordable drugs to treat HIV/AIDS, tuberculosis, and malaria. The program has a budget of $600 million, of which $200 million comes from the EU, $200 million from national research programs, and $200 million from the private sector.[19] Some tropical disease drug and vaccine development is also funded by the U.S. and other national governments as a biodefense expenditure where a disease in question could be used as a biological weapon. An unintended positive consequence of this R&D is that patients in developing countries who are exposed to viruses such as Ebola can benefit from the availability of such drugs or vaccines that otherwise would not have been developed or would have been available much later. Nanotechnology may be an essential element of new drug development technology, diagnostic chips, and perhaps drug delivery systems for orphan and tropical diseases.

1. Discussion

In the pharmaceutical R&D market, the U.S. and Europe compete to attract pharmaceutical companies and foster the development of innovative medicine. Governments on both sides of the Atlantic invest in new drug developments including nanodrug delivery measures and high throughput screening. European governments and the European Commission are also concerned about public acceptance of new technologies including drugs. Investment in cures for tropical diseases is a new priority in funding by the European Commission and national governments of Western countries.

2. Medical Device R&D

The market for medical technologies is far more fragmented than the pharmaceutical market. The European Medical Technology Industry association (EUCOMED) defines medical technology as covering

> a very wide range of products: aids for the disabled, active implantable devices; anaesthetics/respiratory devices; dental devices, electromedical, hospital equipment (hardware), imaging, *in vitro* diagnostic devices, ophthalmic and optical devices, passive implantable devices, single-use (disposable) devices, surgical (reusable) instruments.

The association estimated the world market for medical technology to be $170 billion in 2001. In 2002, spending amounted to $54.8 billion in Europe, $79.3 billion in the U.S., and $20.1 billion in Japan.[22] Market growth is 5 to 8% per year. The market is dominated by 94% small and medium-sized enterprises, each specializing in a niche market. Europe has 9,345 such companies of which >80% are SMEs. The U.S. has 10,000 companies including 98% SMEs. Japan has a total of 1,580 companies.[22]

As noted earlier, governments fund fundamental research. regulate market access, and determine which new technologies will be refunded by public health insurance. Regulation of market access is not as strict or well defined as it is for pharmaceuticals. In Europe, three directives regulate market access of medical devices. These medical devices are defined as follows:

> ...any instrument, apparatus, appliance, material or other article, whether used alone or in combination, including the software necessary for its proper application intended by the manufacturer to be used for human beings for the purpose of: diagnosis, prevention, monitoring, treatment or alleviation of disease; diagnosis, monitoring, treatment, alleviation of or compensation for an injury or handicap; investigation, replacement, or modification of the anatomy or of a physiological process; control of conception; and which does not achieve its principal intended action in or on the human body by pharmacological, immunological or metabolic means, but which may be assisted in its function by such means.[23]

This includes diagnostics and lab-on-a-chip techniques along with dental and medical prostheses and implants involving biomedical nanotechnology. The three relevant European directives are:

1990 Directive on Active Implantable Medical Devices (AIMDD) — Covers medical devices (1) relying for their functioning on sources of electrical energy or power other than those directly generated by the human body or gravity; and (2) intended to be totally or partially introduced, surgically or medically, into the human body or by medical intervention into a natural orifice and expected to remain after the procedure. Typical products covered are pacemakers, diffusion pumps for oncological applications, and cochlear implants.[24]

1998 Directive on *In Vitro* Diagnostic Medical Devices (IVDD) — Covers any medical device that is a (1) reagent or reagent product, (2) calibrator, control material, or kit, (3) instrument, apparatus, equipment, or system, whether used alone or in combination, intended by the manufacturer to be used *in vitro* for the examination of specimens, including blood and tissue donations, derived from the human body, solely or principally for the purpose of providing information (1) concerning a physiological or pathological state or congenital abnormality, (2) to determine the safety and compatibility of devices with potential recipients, or (3) to monitor therapeutic measures. Typical *in vitro* devices covered include reagents for determining pregnancy, reagents for diagnosing AIDS, reagents for determining blood grouping, reagents for determining hepatitis, and specimen receptacles for the containment and preservation of human specimens.[25]

1993 Directive on Medical Devices (MDD) — This directive covers medical devices not subject to the AIMDD or the IVDD including hospital and dentistry equipment, audiometric devices, ophthalmic apparatus, implantable and nonimplantable prostheses, internal and external orthopedic materials, aids for the disabled, and disposable materials.

These three directives set minimum norms and standards for market access of medical devices in the EU member states. The rules are in modular form and different rules apply, depending on the types of risks expected to arise when the devices are used. National governments can set stricter rules, but must allow imports of products from other EU member states if the products were approved in accordance with EU norms. This requirement is known as conformity assessment. Medical devices that include nanotechnology can present medium (classes IIa and IIb) or high (class III) risks. Medium risk class IIa includes dental filling materials that may include nano-structured and other components. The manufacturer is responsible for conformity assessment of product design. The conformity assessment of device production must be backed up by a Notified Body in one of the EU member states. Medium risk class IIb covers medical imaging equipment and other large medical devices including those that insert gas or small particles into the body. The class may include nanoparticles for imaging. The assessments of both design and production must be backed up by Notified Bodies. High risk class III encompasses active and passive implants that may include nanostructured materials or components. Class III is procedurally similar to class IIb, but the manufacturer must also submit a design dossier to the Notified Body for approval.[26]

IV. SOCIETAL BENEFITS AND ETHICAL, LEGAL, AND SOCIAL ASPECTS OF NEW TECHNOLOGY

Nanotechnology is an emerging field in a very early stage of development. Beyond a small circle of experts in academia, industry, and government, few people really understand its potential for biomedical technology. As the earlier chapters in this book demonstrated, biomedical nanotechnology will be integrated into larger scale medical technological systems and products. Biotechnology, ICT, and nano-technology will converge to contribute to these improved systems and products. The answer to the question of what priorities in biomedical nanotechnology research can contribute to societal needs is related closely to the contributions of biotechnology and ICT. The OECD[27] explicitly mentions nanotechnology in an exploration of biotechnology in the fight against infectious diseases. In the OECD's view, nano-technology is one of the potential surveillance techniques for investigating antibiot-ics-resistant tuberculosis and other infectious diseases. Nanotechnology "… is an example of the use of the converging sciences of genetic fingerprinting, nano-technology and automated digital analysis to follow and predict patterns of spread of these difficult-to-treat TB strains" (p. 12).

Other related techniques for improved surveillance of emerging infectious dis-eases include microarrays integrated into an Internet or other network to allow monitoring at a distance and polymerase chain reaction (PCR) techniques to amplify genetic materials. New biotechnologies and related technologies are expected to contribute to improved detection techniques and point-of-care diagnostics including microarrays and biosensors. Vaccines are essential elements for preventing outbreaks of infectious diseases. Trends in converging technologies can contribute to faster development of new vaccines. In a case study of a vaccine for Group B meningococcus, use of new reverse vaccinology led to the development of a vaccine in 18 months. This was not possible earlier, despite 40 years of vaccine research via conventional methods.[27] Converging technologies can also contribute to better treatment of infectious diseases (pp. 19, 22). The report ends with recommendations for an R&D agenda, but it covers applications, not specific technologies. The OECD recommends that both public and private organizations in its member countries should contribute to the R&D agenda.

Other initiatives to set nanotechnology roadmaps and priorities focus more on national needs in countries with specific nanotechnology research programs and on the EU. In this early stage of nanotechnology development at the time of writing this chapter (2003–2004), the debate on ethical legal and social aspects of nanotechnology in general and biomedical nanotechnology in particular is only beginning. A big gap continues to exist between researchers and promoters of nanotechnology development on the one hand and regulators and civil society on the other. Scientists tend not to understand potential societal implications and ethical consequences. Regulators, social scientists, and the general public tend not to understand what nanotechnology is and what related R&D efforts can imply for new drugs and medical device development and possible effects on society. After encountering unexpected public resistance against genetically modified organisms and other biotechnology inventions, especially in Europe, politicians are trying to include technology assessments and discussions about the ethical, legal, and social aspects of new technologies in the early stages of development. Current negative issues related to nanotechnology concern the toxicity of nanoparticles and the "green or grey goo hypothesis" of self-replicating nanobots (see Chapter 7, this volume).

Positive expectations include the generic nature of nanotechnology that can lead to qualitatively better products in many different sectors. The semiconductor industry and pharmaceutical and medical device industries are the sectors where nanotechnology is already well integrated into product and process innovation and R&D. Benefits of nanotechnology are difficult to perceive because nanostructured materials and components serve as parts of larger systems, products, or process technologies. Examples of process technologies are scanning probe techniques applied in industrial quality control and microfluidics or nanofluidics used in R&D and production of small amounts of products in the fine chemicals and pharmaceutical industries. Longer term progress in the pharmaceutical and medical device sectors will lean heavily on today's fundamental nanotechnology research and patenting and the owner of the technology will reap the benefits. In order to achieve healthy and well balanced debates on the pros and cons of nanotechnology, it is essential to explain how nanotechnology can contribute to tomorrow's innovative products.

A. Views of Nanotechnology Experts on Socioeconomic Aspects of Their Work

1. Methods of Coping with Ethical, Legal, and Social Aspects

Constructive technology assessments and science communications receive a lot of attention from researchers and research policy makers in North America and

Europe. Constructive technology assessment implies co-evolution of new technology development and debate on and assessments of potential and desirable societal implications of emerging technologies.[28] The National Nanotechnology Initiative in the U.S., the EU-funded thematic program on nanotechnology, and national nano-technology research programs such as the Dutch NanoNed include funding dedicated to societal aspects of nanotechnology. Several conferences in the U.S. and in Europe have already been dedicated to these aspects. These activities imply more commit-ment from researchers in academia and in industry to consider the societal implica-tions of their research and engage in public discussions about them. For ethically sensitive issues related to pharmaceuticals and biomedical technologies, it is even more relevant to develop a good understanding of these issues from the earliest phases of development.

As the debate between science and society takes shape, governments are begin-ning to stimulate researchers to argue about the benefits to society of their research. In the late 20th century, the main criteria for funding research proposals were scientific quality and economic benefits to the country or region where the funding body was located. Early in the 21st century, funding bodies put more emphasis on the articulation of more general benefits to society, and required scientists to engage in public debates about their work and in discussions with critics of scientific and technological progress.[29] During the preparation of this book, I asked a number of researchers about the societal contexts in which their research took place (see below). Many scientists engaged in biomedical nanotechnology research are still working mainly in the contexts of other researchers, clients in industry and SMEs, and government funding bodies. The broader public debate involving stakeholders such as medical professionals, patients, consumers, and outsiders is not directly relevant to the daily work of most researchers. Medical professionals frequently collaborate in applied research projects.

2. Researchers' Opinions about Societal Aspects of Their Research

a. Nanodrug Delivery

A professor specializes in nanoparticles as carriers for drug delivery and drug targeting and transport of drugs across the blood–brain barrier. Advisory committees of his projects include representatives of large industries, SMEs, funding agencies, governments, and medical professionals. The professor's research is at the preclinical testing phase and time to market is estimated as about 5 years. The end users of the results are pharmaceutical companies and SMEs. The professor sets his own research priorities. He selected brain tumors, multiple sclerosis, and gene therapy as research priorities because "no effective therapy exists against brain tumors as well as multiple sclerosis, and we made already good progress in the first field mainly due to serendipity." He has a pragmatic attitude about setting priorities: "Any area which is funded. Money is the essence of research."

Two Ph.D. students are involved in longer term research (more than 10 years to market) on nanodrug delivery. The first works on pharmaceutical technology and delivery, the second on microencapsulation, nanomaterials as host molecules, and a kinetic model of drug release via nanomaterials. Projects for both students include user representatives in an advisory group — the first from a large industrial concern and the second from the government. Pharmaceutical companies are among the end users of the products. One student works also to benefit patients and consumers; the other works also for the eventual benefit of medical technology manufacturers. The motivation for the research of the first student is socioeconomically oriented: improving existing therapies along with the expectation of a huge market. The other is more technology oriented: "It helps to understand the behavior of nanoparticles and eventually to be able to control them in such a way as the application needs."

The first student believes that the top priorities for biomedical nanotechnology research worldwide are cancer, HIV, diagnostics, tissue engineering, and molecular machines. According to him, the patients treated determine his priorities. The other student has no opinions on these strategic issues. The first student has no other preferred priorities for research, whereas the second would like to work on surface modification. The first student sees three issues that must be addressed to improve the societal impact of his research: "There are too many law restrictions, patent filing takes too long in Germany, and there is a need for more government funding." The second student is skeptical about the possibility of taking measures to improve the societal impact of his research: "The problem is in particular that research in the field of biomedical products takes years and years because of the need of many tests before being commercial."

b. Diagnostics and Drug Discovery

A product manager works for an SME that designs nanostructured surfaces for handling small liquid volumes for bioanalytics and diagnostics. The applications of the R&D in this company are diagnostics, drug discovery, and digital lab-on-a-chip platforms. Outside users are not involved in monitoring the progress of the research. The time to market is 3 months to 2 years. The research covers pilot production, optimization, and proof of principle of biochips. The end users of the company's products are contract research organizations and pharmaceutical companies. The motivation for these activities is economic: "Genomics and proteomics are future technologies that become more and more important. Analysis and handling of very small amounts of liquids with high throughput potential and high sensitivity are the big challenges. It is an area with very high market potential and a wide variety of possible applications."

3. Discussion

Technology assessment is beginning to be included in nanotechnology research projects under pressure of public opinion, especially in the EU. National and EU policy makers are concerned about public acceptance for innovation in general and nanotechnology in particular and are asking for the inclusion of social scientists and science communication activities in the research they fund. Some leading researchers have also taken initiatives to include technology assessment activities in their research projects, for example, researchers in the Dutch NanoImpuls and NanoNed programs have taken the initiative to include constructive technology assessment.

B. Main Issues

Most researchers who work in laboratories are not accustomed to thinking about the societal implications of their work and lack the skills and knowledge to effectively address the concerns of the general public. It is unclear where the emerging debate about ethics and societal aspects of research will lead. This section will discuss some possible issues for debate. Some of the expected societal benefits and ethical, legal, and social issues related to pharmaceutical and medical technology R&D in general and biomedical nanotechnology in particular are also covered.

1. Opportunities

Reducing side effects of treatments, especially chemotherapy — This potential benefit is usually cited as a basis for improved delivery of anticancer drugs. The benefit may also apply in other situations such as targeted drug delivery to help prevent side effects arising from use of several drugs at the same time. Drug delivery systems for treating hepatitis C and intestinal infections are also under development. The potential use of magnetic particles to target cancerous tissues and application of an electromagnetic field to destroy cancer cells by heating them locally are other possible adjunct methods that may improve the effects of chemotherapy even though they are not effective as stand-alone cures.

Faster and more intelligent selection of active drug compounds — This is a general trend in high throughput screening and controlled design of drug molecules.

Improving quality of life of disabled patients with prostheses and implants — Certain types of prostheses and implants may actually grow into the body and impose fewer risks of rejection. Nanotechnology or tissue engineering may eventually help cure paralysis, brain damage, and diseases of the central nervous system by restoring neurons.

More timely diagnosis of diseases to improve chances for successful treatment — This potential benefit is often cited. Then President Bill Clinton mentioned it in his announcement of the National Nanotechnology Initiative in 2000. Nanotechnology is one member of a group of converging technologies that will help develop new diagnostic tests for cancer and alternatives to replace current scanning techniques such as x-rays and nuclear magnetic resonance.

Monitoring nuclear, biological, and chemical warfare agents and other toxins — See Chapter 5, this volume, for a discussion of the use of nanotechnology to monitor toxic materials in the environment and in the human body.

2. Challenges

Safety of new drugs, implants, and artificial organs — Medical devices and the active components of new drugs are checked for safety before they are allowed on the market. New drugs must pass a series of preclinical and clinical trials before government approval is granted. After market introduction, regulators monitor for unexpected health hazards. The European Commission has noted that there is room for improvement and harmonization of the national monitoring practices of EU member states.[20] A new issue arising from advances in nanotechnology is the disappearance of the boundary between pharmaceuticals and medical devices. For example, should implants that have drug-eluting coatings be treated as pharmaceuticals or medical devices? Can a medical device with such a coating legally be put on the market without the extensive testing required for pharmaceuticals? Eliminating the testing could lead to unforeseen risks. The issue may require adaptation of existing regulations.

Costs of health care and biomedical technology — These costs must be controlled. Social and technological trends over the past years imply rising costs of health care in western countries and point to the need for choices in insurance coverage for new products and in funding of R&D for new technologies. Debates about how to make such choices continue. Some authors feel that a fair evaluation of insurance coverage should compare both established technologies and new technologies. The main argument is that new technologies are expected to be superior to older ones. To accommodate this view, the European Commission[20] intends to implement the concept of "relative effectiveness" in the evaluation of health technologies in future policies, but only after the medicines are in the market. The main criteria for evaluating new drugs and medical devices remain safety, quality, and efficacy. "Relative effectiveness, as applied to health care technologies such as medicines, has two components: the added therapeutic value (ATV) of a medicine (its clinical effectiveness compared to other treatments), and its cost effectiveness, which builds on ATV and brings cost considerations into the comparison."[20] The implementation of this concept in European policies could speed up market access of biomedical nanotechnology and other medical technologies in Europe and also contribute to better and more cost-effective treatments. New drugs and medical devices would then have a better chance of competing with obsolete but well established products and treatments.

Privacy issues related to genetic information — Robert Rizzo[30] predicts that genetic tests will be used routinely by medical doctors to determine genetic dispositions before they prescribe treatments for patients. In my view, this socioeconomic scenario can be aided by nanotechnology developments of biochips and diagnostic tests (see Chapter 4, this volume). Rizzo argues that genetic information, if stored, could be accessed by others including insurance companies that may be tempted to use such personal information to decide whether to issue health insurance to the individual concerned. Rizzo notes that this would be hard to prevent through legislation because of the extent of commercial interests involved in market-dominated health insurance systems like those in the U.S. He therefore expects public compulsory health insurance systems to be predominant in the future. In my opinion, his argument about the causal chain between genetic testing and the predominant system of health insurance goes against the present trend in market-based health insurance systems based in part on a general globalization trend in which formerly public sectors such as transport infrastructure, energy supply, and health care are becoming more privatized. It is not likely that the emergence of one new technology, however generic it may be, will reverse this trend.

Access to affordable health care and biomedical technology (including orphan drugs) — The lack of access to new biomedical technologies by uninsured people, people in developing countries, and victims of orphan diseases in developed countries is getting more attention from policy makers. Among the reasons for dedicating specific funds are the catastrophic impacts of HIV/AIDS and other infectious diseases in sub-Saharan Africa and the ensuing human tragedies. Biomedical nanotechnology research could help alleviate these problems if researchers apply for funding dedicated to these diseases and collaborate with researchers and companies in developing countries on international R&D projects. These types of health problems cannot be solved by technology alone.

Democratic choices in new technology development and health insurance coverage — As noted earlier in this chapter, different types of stakeholders are involved in the markets for health care and for biomedical R&D. Patients, consumers, and medical professionals are usually excluded from R&D decision making and outsiders are excluded from both markets. This implies that the needs of patients and consumers are considered in the late phases of technology development when many priorities for R&D funding have already been decided. A strong push in new technology development can lead to inefficiency if the technology does not address genuine health care priorities and can lead to unexpected public debate about the ethics and risks attached to new biomedical technology.

Standardization — No standards yet exist for biomedical nanotechnology, although they are not yet needed because most of the nanotechnological research is still fundamental. In Europe, the CE norms for medical devices apply also to relevant nanotechnology and products that include nanotechnology. The standard regulations for approval of new pharmaceuticals apply to nanodrug delivery systems. EUCOMED, the European medical technology industry association, wants to start a debate on the need for new norms and standards in collaboration with the international and European standardization bodies (ISO, CEN/CENELEC, etc.). CEN has been trying to organize more future oriented conormative and prenormative discussions about new technologies. In 2002, it organized a discussion on standardizing nanotechnology instruments for research and industrial quality control. These laboratory instruments can be used for drug discovery and diagnostics and include scanning probes and cantilever diagnostics that are difficult to calibrate. Therefore they cannot be used as quantitative measurement instruments. These standardization organizations are set up and maintained by industrial companies on a voluntary basis. The standardization research is carried out in government laboratories and funded by governments. Nongovernment organizations such as consumer groups are also represented in the standardization bodies.

Another issue related to norms and standards for biomedical nanotechnology concerns classification. Does biomedical nanotechnology fit into the pharmaceutical or medical device category or should a new category be devised to cover it? Drugeluting stents serve as examples that illustrate this dilemma. A stent is an implanted medical device covered with a nanocoating that delivers drugs (pharmaceuticals) to the circulatory system to prevent thromboses and infections. Should the stents be regulated by the medical device directive or by the pharmaceutical directive, or should a new category with norms and standards be created? Early in 2004, CEN formed a new committee to discuss the needs for new nanostandards.

Ethical issues related to cybernetic organisms (cyborgs) and other bioethics issues — Kevin Warwick[31] is currently experimenting with a neuroprosthetic device in his arm that integrates him and a computer network via a wireless connection. The implant provides him with remote control of the computer and also enables the computer or people operating the computer to exert some control over Warwick's movements and even his feelings. This concrete example of a living cyborg (cybernetic organism — a combination of human and machine) renders ethical and privacy issues a lot more urgent than they were earlier when cyborgs were only discussed in science fiction movies. Most current research on active implants relates to the development and application of microsystems technology (MEMS). Nanotechnology is beginning to be included in MEMS devices as biocompatible or drug-eluting coatings, in small chip-integrated batteries, or in surface treatments to improve connections between implanted devices and surrounding tissues.[13]

Risks of abuse of the new technologies — Chapters 5 and 7 of this volume discuss biodefense issues, the possible misuse of new technology for purposes of warfare, terrorism, or crime, and the potential for jeopardizing the public health and the environment.

3. Beneficiaries and Patients

If successful, biomedical nanotechnology will benefit the pharmaceutical and medical device industries and life science and biomedical SMEs through higher earnings achieved from products that contain nanostructured materials and nano-components. Patients who are disabled or suffer from cancer, central nervous system diseases, HIV/AIDS, and other infectious diseases are likely to benefit from better cures and enjoy longer and healthier lives. The possible disadvantages may arise from (1) potential harmful effects of nanoparticles on patient health or (2) unexpected side effects of drugs, treatments, and medical devices. Patients in the former case are generally not beneficiaries of biomedical nanotechnology development; patients in the latter case are.

Other sufferers of negative consequences of biomedical nanotechnology without profiting from the benefits are opponents of medicalization of disabled individuals who feel that implanted devices make disabled people depend on technology to lead normal lives. This group favors greater public acceptance of disabled people as valued members of society. The consequences to taxpayers for incorporating bio-medical nanotechnology into drugs and medical devices are uncertain. Such innovations may increase the costs of treatment, but may also decrease related costs for insurance and other types of care.

Finally, outsiders will not benefit from biomedical nanotechnology, but they will not suffer negative consequences directly. Indirectly, they may face decreased life expectancy compared to those who will directly benefit. Those who suffer from negative effects of nanoparticles and the opponents of medicalization of the disabled can be expected to be the main protagonists if opposition to biomedical nanotechnology emerges. The first group will not suffer from direct effects of biomedical nanotechnology; they will be affected by airborne particles or those contained in products implanted in the body or taken up through the lungs, intestinal tract, or possibly the skin. Perhaps the concerns of the opponents of medicalization of the disabled should be considered in the designs of new biomedical technologies to minimize any infringements on freedom of movement and ensure that implants and prostheses will be effective as long as their bearers live. That would reduce the need for surgery, which is still not a risk-free procedure. However, in democratic societies, even scientists must cope with different opinions of people who may or may not welcome new technologies.

V. ADDRESSING ETHICAL, LEGAL, AND SOCIAL CONCERNS

A. Regulation

The Food & Drug Administration (FDA) regulates new drugs and medical devices in the U.S. The Shonin is the regulatory body in Japan. In the EU, market access to new drugs has been regulated by the European Medicines Evaluation Agency (EMEA) since 1995. The FDA is already organizing conferences and discussions concerning regulatory aspects of medical nanotechnology. Before a new drug is allowed on the market in Europe, Japan, or the U.S., the pharmaceutical company developing it must first prove that it works and is safe and nontoxic. The company must conduct standardized preclinical trials and three phases of clinical trials of the product. After each phase, it must submit files and test results to the FDA or other regulatory authority. The FDA has the strictest rules. For medical devices, a similar procedure applies before a device is allowed on the market. In the EU, a CE designation is required to show that a product meets safety and performance requirements in accordance with the three directives discussed above.

Nanotechnology may require the development of new regulations since it erodes the boundary between pharmaceuticals and medical devices.

B. Health Technology Assessment

Medical or health technology assessment is a well established instrument for evaluating possible negative side effects or health risks of new drugs or medical devices. The assessment includes a cost–benefit evaluation of these technologies. Medical technology assessment is defined as "the evaluation or testing of a technology for safety and efficacy. In a broader sense, it is a process of policy research that examines the short- and long-term consequences of individual medical technologies and thereby becomes the source of information needed by policy makers in formulating regulations and legislation, by industry in developing products, by health professionals in treating and serving patients, and by consumers in making personal health decisions."[32] Health technology assessment is well integrated in the market approval procedure for pharmaceuticals. Assessment requirements for medical devices are still under development.

EUCOMED issued a position paper on health technology assessment of medical devices in Europe (2001). It is in favor of a pragmatic approach to assessment of medical devices because the average time for medical device innovations is 18 to 24 months — a far shorter time than that required for assessment of a drug. Medical devices are continuously subject to evolutionary innovation and products already on the market are continually improved.[33]

Health technology assessment professionals are discussing the needs and possibilities for integrating technology assessment activities in earlier phases of the development of new medical technologies in the hope of including other criteria in evaluations of proposals to develop new medical technologies. For example, during the development of ultrasound techniques, decisions to fund one type of equipment rather than another in early stages of development determined the product that eventually reached the market. Ultrasound is useful for illustrating the shapes and placements of organs in the body; it is less valuable for distinguishing healthy and cancerous tissues. If medical professionals or representatives of patient associations had been involved in the proposal evaluation, the technology may have been different. The decisions related to ultrasound were made by physics-oriented evaluators.[34]

The assessment of nanotechnology is still in an early stage. At the time of writing of this chapter, results of studies conducted in Germany and Switzerland between 2001 and 2003 have been published and new studies are in progress. Conferences have been organized; critical nongovernment organizations such as the ETC group and Greenpeace and critical scientists have initiated public debates, particularly in the United Kingdom where Prince Charles sparked public interest.

The Swiss Centre for Technology Assessment (TA-Swiss) was the first parliamentary organization to engage in an assessment of nanotechnology, particularly the medical, social, and ethical aspects of nanotechnology in medicine and the likely impact of medical nanotechnology on society (2001–2003). A panel of experts expects that by 2020 medical nanotechnology will be likely to contribute to therapies for cancer, followed by therapies to treat bacterial infections and cardiovascular diseases. Medical nanotechnology is less likely to play a role in treating metabolic and autoimmune diseases and viral infections or curing Alzheimer's disease.

The authors of the TA-Swiss assessment study propose to set up a competent international body to monitor the development of nanotechnology, identify possible risks early on, and ensure that these risks are mitigated. This body should also stimulate communications between scientists and society, organize health campaigns to make citizens aware of the potential of medical nanotechnology, and analyze the need for new forms of health counseling created by the emergence of medical nanotechnology.[35]

The Technology Assessment Bureau of the German federal parliament engaged in a more general study that included material about the impact of nanotechnology on legislative requirements. The study included an evaluation of applications of nanotechnology in life sciences and their market potentials based on a literature review. As of 2002, the review noted that microscopic research of biological objects and cosmetics had substantial market potential, followed by implants and medical instruments that include nanocoatings or surface treatments. The first products to include nanotechnology will be diagnostics, foods, and biomimetics devices.

Nanotechnology market studies for 2020 predict dominant shares for implants and medical instruments and substantial market shares for microscopy, diagnostics, drug delivery systems, foods, and cosmetics.[36]

1. Need for New Legislation

New legislation will be required only if the technical properties of nano-technology and the interactions of nanotechnological devices with humans or the environment are not sufficiently covered by existing legislation, perhaps because existing definitions are not adequate. The current legislative definitions appear broad enough to cover both pharmaceutical and medical device applications of nano-technology. A further need for legislation may arise if nanotechnology innovation leads to incremental product improvement and the improved product does not meet existing regulative controls. This appears to be the case with nanoparticles contained in products such as sun block preparations. The existing creams contain the same materials, but the particle size may lead to unexpected health risks (see Chapter 7, this volume). The debate about regulation of nanotechnology is only beginning in 2004 and the outcome is unclear.[37]

C. Participatory Technology Assessment

Several methods to accomplish participatory technology assessment bring together different stakeholders to discuss potential societal (ethical, legal, and social) consequences and potential benefits. One of these methods is the consensus confer-ence at which a panel of lay people are informed about a new technology, discuss the technology, and produce a consensus on relevant and desired policy measures to guide the development and applications of the technology. The results are pub-lished and made available to parliament members or other decision makers. Con-sensus conferences were invented in Denmark and are also popular in The Nether-lands. Whether such forms of direct democracy are useful depends on the type of national government. Other methods involve scenarios or future workshops to explore potential of developments. Delphi studies, in which a group of experts are asked to estimate the time in which a technology is likely to enter the market, are also used. In the first round of Delphi studies, experts make their estimates individ-ually; in the second round, all the experts' estimates are shared and the group is given a chance to change its estimate.

One problem with consensus conferences is whether they are representative of public opinion in a country as a whole. Another problem is more relevant for nanotechnology: in the early stages of technology development, it is difficult to imagine what future products may reach the market. It is even more difficult to foresee potential societal implications and future public opinion. Finally, it is not easy to motivate people to spend time discussing a new technology unless it is surrounded by a political issue that stimulates interest.

In the constructive technology assessment approach followed by the NanoNed program in The Netherlands (2003–2007), dialogue workshops that include research-ers and lay people are held in parallel to nanotechnology R&D projects. The aim is not so much to find issues for policy makers, but to stimulate better quality and more socially relevant developments by making researchers aware of public attitudes toward their technology and enable them to better explain their work and its potential implications for society. The U.S. Congress passed a nanotechnology bill late in 2003 that covers funding for research on societal aspects of nanotechnology. The total budget for all research is $3.7 billion over 4 years but it provides no fixed budget for studying societal aspects. A presidential advisory committee will report biannually whether societal aspects are adequately addressed.[38]

Other methods in participatory technology assessment include organizing hearings in parliaments by government administrators or by national or EU parliamentary technology assessment organizations. Nongovernmental organizations such as the ETC group and the European Nanobusiness Association have also organized seminars to raise the awareness of the parliament members about societal consequences or benefits of nanotechnology.

D. Technology Forcing

Technology assessment specialists have developed methods for technology forcing — an approach that attempts to set targets for new technology development and actively stimulates the realization of these targets. A well known example is Moore's law in semiconductor miniaturization which is technological and economical in nature. Technology forcing requires roadmaps that include technological, economic, and societal aims. Two sensitive issues related to technology forcing are (1) deciding which industry or government organizations set the priorities and (2) determining who is excluded from setting the priorities and thus will be less likely to agree to the introductions of new products and benefit from the outcomes.

Other methods such as backcasting attempt to set societal goals and calculate backward the technological developments and policy measures required to achieve targets. Backcasting was devised to deal with sustainable development. For example, policy makers can set strategic goals such as reducing CO_2 emissions by 8% between 2008 and 2012, compared to 1990 levels (EU target in conformity with the Kyoto Protocol on climate change). They then develop quantified scenarios of the steps required to meet the targets.

The targets set by the WHO for biomedical nanotechnology cited early in this chapter could eventually benefit cardiovascular disease or cancer victims by the year 2020, after which backcasting techniques could be used to devise a strategy to achieve the health care and resource goals projected for 2020. The result should be a general roadmap, including but not limited to setting the necessary priorities for development.

Another goal for biomedical nanotechnology could involve addressing two of the eight Millennium Development Goals formulated by the UN General Assembly (see introduction to this chapter) — a focus on major diseases such as AIDS and malaria and the formation of an international collaboration for development. The latter would imply setting up research networks including researchers from the northern and southern hemispheres along with investments in infrastructures such as high speed telecommunication links and high quality research facilities in developing countries.

Roadmapping exercises related to nanotechnology are already included in the EU's Sixth Framework Program of Research and Technology Development, but they are mainly technological and economic in scope. In the U.S., 12 grand challenges for nanotechnology research are already in place. They are also mainly technological in scope (e.g., nuclear, biological, and chemical weapon-sensing devices).

VI. CONCLUSIONS

The main health risks according to the WHO relate mainly to poverty in the southern hemisphere or the unhealthy behavior (e.g., obesity) of populations in rich countries. While nanotechnology development is not directly relevant to those problems, infectious diseases, cancer, cardiac diseases, and disabilities claim millions of victims each year. Nanotechnology is an emerging area of the pharmaceutical and medical device industries and products based on it will enter Western markets in the coming decades. Nanotechnology is likely to alleviate suffering and contribute to healthy life expectancies of many people.

Several stakeholder groups in our global society potentially have interests in the development of biomedical applications of nanotechnology. Certain groups including researchers, industrialists, and governments are actively engaged in choosing what nanotechnology will eventually contribute to health care. Other groups such as medical professionals and patients will benefit at a later stage. Still others are unlikely to benefit because they will not be able to afford nanotechnological pharmaceuticals or medical devices, and another group may encounter negative consequences of medical nanotechnology and nanoparticles if potential health risks are not identified and remedied early enough.

A number of global trends in health care systems and economics are likely to influence the development of nanotechnology and may in turn be influenced by the emergence of biomedical applications of nanotechnology. The UN General Assembly set targets for the organization of an international knowledge economy by 2015. The WHO forecasted trends in the incidence of major lethal and disabling diseases through 2020 and set targets for policymakers to reduce the numbers of victims. Researchers, industrialists, and government funding agencies could use these targets in their research and funding strategies to force the development of biomedical applications of nanotechnology and work toward fulfilling the world's needs for better health care in the coming decades. Cardiac disease accounts for 16.5 million victims each year. The WHO expects cardiovascular disease to be the leading cause of death in developing countries in 2010 and proposed several measures to prevent this scenario from becoming a reality. The measures include healthy living campaigns, improved monitoring networks to allow early identification of patients, and distribution of low-cost medications. Nanotechnology applications in diagnostics and drug development could contribute to progress in this area. The use of nano-structured materials and coatings for active and passive implants such as stents and pacemakers would also help reduce the number of victims of cardiovascular disease. In 2000, 10 million people developed new cancers and 6.2 million people died of cancer. By 2020, the WHO expects 15 million people to develop new cancers every year; it aims to prevent a third of those cases and cure another third. Nanotechnology may be able to contribute to new cures for cancer via diagnostics and drug delivery systems. Infectious diseases including HIV/AIDS, malaria, tuberculosis, and others produce millions of victims annually and the number may grow as a result of attacks with biological weapons. Nanotechnology may contribute to fighting these diseases through better diagnostics and sensors and discovery of new drugs.

Other global trends relevant to world health are the aging populations in Western countries and the increased occurrence of "welfare diseases" (heart attacks and strokes) in richer developing (newly industrialized) countries. Rising medical costs have led to new debates on priorities for health care insurance and funding. Whether innovation in pharmaceuticals and medical devices leads to an increase in health care costs or produces the opposite effect is still an open question.

The markets for biomedical nanotechnology R&D are the pharmaceutical industry, the medical device sector, and possibly a combination of both. Nanotechnology is already important in the pharmaceutical industry as an aid to discovery and a component of new drug delivery vehicles. In the medical device field, nanotechnology has more of a long-term role, particularly in microsystems technology. Nanostructured materials are already included, for example, in pacemaker electrodes. Nanotechnology may blur the boundary between pharmaceuticals and medical devices and necessitate new regulations to determine market access of new products. It may also lead to a reorganization of what presently constitute separate markets.

Since about 2000, policy makers, scientists, and industrialists have talked about stimulating debates between science and society. In 2003, the discussions intensified after criticisms by the Canadian nongovernmental organization known as the ETC group and by Prince Charles of the U.K. However, in practice, any debate about the uses of biomedical and other forms of nanotechnology is still embryonic. Some initiatives have been set in motion, especially for constructive technology assessment projects and studies of parliamentary technology assessment organizations and other social scientists. Initiating attempts at technology forcing in early stages of technology development is still very rare. At these stages, priorities are set for longer term research. This means the decisions taken then are important determinants of the properties of the technologies which will enter the market in the future.

Because most victims of severe infectious diseases, particularly HIV/AIDS, live in developing countries, research should be targeted on finding cures for such diseases. Researchers working in facilities in non-Western countries should be included more often in international projects; funding should be made available to stimulate spin-offs of start-up companies that develop technology into marketable products. The internet and ICT could facilitate the emergence of real global knowledge about biomedical nanotechnology. The efforts of Mihail Roco, coordinator of the U.S. National Nanotechnology Initiative, to promote international initiatives should not be limited to rich countries if nanotechnology is to exert a real impact on world health. The International Nanotechnology Initiative could very well serve as the motor for achieving the Millennium Development Goals of the UN.

REFERENCES

1. United Nations General Assembly. Resolution 55/2, Millennium Declaration. New York, 2000.
2. Prentice, T., Ed. *World Health Report: Reducing Risks, Promoting Healthy Life.* World Health Organization, Geneva, 2002.
3. World Health Organization. Malaria is alive and well and killing more than 3000 African children every day: WHO and UNICEF call for urgent increased effort to roll back malaria. Press Release. Nairobi, April 25, April 2003. www.who.int
4. World Health Organization. Fact Sheet 180: Variant Creutzfeldt-Jakob Disease. Geneva, November 2002. www.who.int
5. Hüsing, B., Gaisser, S., Zimmer, R., and Engels, E.M. Cellular Xenotransplantation. TA-Swiss, Centre for Technology Assessment, Berne, 2001.
6. Stewart, B.W. and Kleihues, P. *World Cancer Report.* WHO IARC nonserial publication. Geneva, 2003.
7. European Agency for the Evaluation of Medicinal Products. Orphan Medicinal Product Designation in the European Union. London, 2003. http://www.emea.eu.int/pdfs/human/comp/leaflet/661801En.pdf
8. World Health Organization. Fact Sheet 213: Blindness. Global Initiative for the Elimination of Avoidable Blindness. Vision 2020. Geneva, February 2000. www.who.int
9. World Health Organization. Fact Sheet 214: Blindness. Control of Major Blinding Diseases and Disorders. Vision 2020. Geneva, February 2000. www.who.int
10. World Health Organization. Fact Sheet 215: Blindness. Human Resource Development. Vision 2020. Geneva, February 2000. www.who.int
11. World Health Organization. Fact Sheet 216: Blindness. Infrastructure and Appropriate Technology. Vision 2020. Geneva, February 2002. www.who.int
12. Dobelle Institute, Portugal. Website: www.dobelle.com, posted 2002.
13. Nanoforum Consortium. Nanotechnology and Its Implications for the Health of the EU Citizen. Nanoforum, Stirling, 2003. Published online at www.nanoforum.org
14. Cutler, D.M. and McClellan, M. Is technological change in medicine worth it? *Health Affairs* 20, 11–29, 2001.
15. Cutler, D.M., McClellan, M., Newhouse, J.P., and Remler, D. Are medical prices declining? Evidence from heart attack treatments. *Q. J. Econ.* 113:4, 991–1024, 1998.
16. Docteur, E., Suppanes, H., and Woo, J. The United States health system: an assessment and prospective directions for reform. OECD Economics Department Working Paper 350. Paris, February 27, 2003. www.oecd.org/eco
17. Imai, Y. Health care reform in Japan. OECD Economics Department Working Paper 321. Paris, February 12, 2002.
18. Pharmaceutical Research and Manufacturers of America. Annual Report, 2003. www.phrma.org
19. Busquin, P. Europe and Africa together in the fight against poverty-related diseases. Speech, September 1, 2003, Johannesburg.
20. European Commission. Communication from the Commission to the Council, the European Parliament, the Economic and Social Committee, and the Committee of the Regions. A Stronger European-Based Pharmaceutical Industry for the Benefit of the Patient: A Call for Action. European Commission, Brussels, July 1, 2003.

21. Council of the European Union. Internal Market Council Conclusions. Brussels, May 18, 1998.
22. EUCOMED. Industry Profile, 2003. Brussels, 2003. www.eucomed.org
23. Council of the European Communities. Council Directive 93/42/EEC Concerning Medical Devices, June 14, 1993. *Off. J.* L169, 12-17-1993, p. 1. Office for Official Publications of the European Communities, Brussels, 1993
24. Council of the European Communities. Council Directive 90/385/EEC on the Approximation of the Laws of the Member States Relating to Active Implantable Medical Devices, June 20, 1990. *Off. J.* L189, 20/07/1990, p. 17. Office for Official Publications of the European Communities, Brussels, 1990.
25. Council of the European Communities. Directive 98/79/EC of the European Parliament and of the Council on *In Vitro* Diagnostic Medical Devices, October 27, 1998. *Off. J.* L331, 07/12/1998, p. 1. Office for Official Publications of the European Communities, Brussels, 1998.
26. Stegg, H., Whitelegg, C., and Thumm, N. The Impact of Single Market Regulation on Innovation: Regulatory Reform and Experiences of Firms in the Medical Device Industry. Publication EUR 19734 EN. Institute for Prospective Technological Studies, Joint Research Centre, European Commission. 2000. www.jrc.es
27. OECD Biotechnology Unit. Biotechnology and Sustainability. The Fight against Infectious Diseases. Paris, 2003. www.oecd.org
28. Rip, A., Misa, T.J., and Schott, J., Eds. *Managing Technology in Society: The Approach of Constructive Technology Assessment.* London, Pinter Publishers, 1995.
29. Malsch, I. Nanotechnologie en Cult (Nanotechnology and Cult.). Paper presented at Studium Generale Maastricht, April 8, 2003 (in Dutch).
30. Rizzo, R.F. Safeguarding genetic information: privacy, confidentiality, and security, in Davis, J.B., Ed., *The Social Economics of Health Care*, London, Routledge, 2001, p. 257.
31. Warwick, K. Identity and privacy issues raised by biomedical implants, in Kyriakou, E., Ed., *IPTS Report 67*, Institute of Prospective Technological Studies, Seville, September 2002.
32. U.S. Office of Technology Assessment, Strategies for Medical Technology Assessment. Washington, D.C. September 1982, p 3.
33. EUCOMED. Health Technology Assessment for Medical Devices in Europe: What Has to be Considered? Position Paper, June 7, 2001. www.eucomed.org
34. Koch, E.G. Why the development process should be part of medical technology assessment: examples from the development of medical ultrasound, in Rip, A., Misa, T.J., and Schott, J., Eds. *Managing Technology in Society: The Approach of Constructive Technology Assessment.* London, Pinter Publishers, 1995, p. 231.
35. Baumgartner, W., Jäckli, B., Schmithüsen, B., Weber, F., Borrer, C., Bucher, C., and Hausmann, M. Nanotechnology in Medicine. TA-Swiss, Centre for Technology Assessment, Berne, 2001. www.ta-swiss.ch
36. Paschen, H., Coenen, C., Fleischer, T., Grünwald, R., Oertel, D., and Revermann, C. TA Project Nanotechnologie: Endbericht. Arbeitsbericht 92. Technology Assessment Bureau of the German Parliament, Berlin, 2003.
37. Nanoforum Consortium. Benefits, Risks, Ethical, Legal, and Social Aspects of Nanotechnology. Nanoforum. 2004. www.nanoforum.org
38. Marburger, J. Speech presented at Workshop on Societal Implications of Nanoscience and Nanotechnology. National Science Foundation. Arlington, VA, December 3, 2003. www.nano.gov.

CHAPTER 7

Potential Risks and Remedies

Emmanuelle Schuler

CONTENTS

I. INTRODUCTION

As the industry around nanomedical devices and drugs emerges, many ponder whether its progress may be hampered by the public skepticism surrounding genetically modified organisms and stem cell research. Under the pressure of the media and some special interest groups, governments in both the United States and Europe have started to conduct studies on the safety of nanomaterials and risks it may impose on human health and the environment. Regulatory processes are expected to take shape soon.

The goal of this chapter is to identify risks associated with biomedical nanotechnology, review the scientific state of knowledge, and provide an overview on how the various stakeholders may respond as nanotechnology matures to large scale production and commercialization. This chapter concludes by suggesting different potential paths for regulation.

Nanotechnology is a field in constant motion and rarely a week passes when the safety and regulation of nanotechnology are not under discussion. As data develop and discussions of risks and safety continue, certain aspects of the issues presented here are likely to gain increased attention over time while others may garner less interest. Nevertheless, this chapter aims to present some elements for future reflection about risk assessment and policies surrounding the production and use of nanomaterials in biomedical applications.

II. NANOMATERIALS IN THE HUMAN BODY

The emphasis is on "polluting" nanosized residues that may end up inside the human body as a result of side effects caused by medical devices or drugs administered for therapeutic purposes. Polluting nanosized residues are nanomaterials or combinations of them that are not intended to reside inside the human body and may have adverse effects on human health. One way to identify these potential polluting nanosized residues is to examine the various biomedical applications of nanomaterials from which they may originate.

Nanomaterials can be utilized in the form of metallic alloys or composite materials of increased biocompatibility and durability in implants; in the form of nanofibers for bone regeneration; in the form of nanoparticles used for diagnostic purposes; and as coloring agents in cosmetics, pharmaceuticals, and paints. The identification of potential health risks arising from nanosized residues is speculative at the moment because most of the applications of nanomaterials to medicine are at early stages of development. The next section discusses examples of biomedical applications.

A. Implants

Nanostructured materials may find a number of applications in implants, primarily to reduce wear and extend performance. Nanomaterials in implants may be tailored to specific parts of the body to adapt to specific needs, offering unique treatment alternatives to existing implants.[1,2] However, some researchers foresee that loose particles from implants made of composites containing nanomaterials may unintentionally lodge within the body as the implant material wears down over time. Whether the body will eliminate these loose nanomaterials or they will cluster into specific organs and become sources of future health problems is still uncertain.

B. Bone Regeneration

Nanostructured materials have been used to heal broken bones. Small pieces of nanostructured calcium phosphate cement measuring 30 nm in thickness by 60 nm in width were used successfully to aid the growth of natural bones after the removal of tumors.[3,4] Such regeneration techniques may replace conventional bone grafting that involves using part of another bone to repair a fracture or fill a cavity. The nanomaterial cement tested in bones may also have some utility in dentistry. The advantage of using nanosized calcium–phosphorus material is that it is biodegradable and disintegrates after about 6 months, unlike conventional therapies, thus reducing the risk of infection. The bone cement nanotechnology tested on patients was recently approved by China's Food and Drug Administration.

Another approach to bone regeneration is a nanoscale molecular scaffolding that mimics the basic structure of natural bone.[5] This synthetic scaffolding is composed of organic nanofibers about 8 nm in diameter and several micrometers in length. These fibers act as pillars for the growth of hydroxyapatite crystals in a way that reconstitutes the original structure of the bone. Nanofibers in gel form are initially injected into the bone cavity that needs repair. The gel hardens as mineralization takes place, producing a material that mimics the original structure of the natural bone. Potential side effects of these nanostructure materials have not been reported.

C. Diagnosis and Treatment of Diseases

Nanoparticles such as semiconductor quantum dots are promising fluorescent probes for cellular imaging.[6,7] Quantum dots are typically nanoparticles made of cores (5 to 10 nm in diameter) of cadmium sulfide, cadmium selenide, or cadmium telluride coated with organic molecules. Quantum dots offer more advantages than conventional organic dyes because their optical spectra are well defined and can be tuned by appropriately varying the chemical compositions and sizes of the cores.

For example, cadmium selenide emits in the ultraviolet and blue part of the spectrum; cadmium sulfide emits visible light; and cadmium telluride emits in the far red or infrared region. Tuning the size of the core makes it possible to obtain a very well defined wavelength. This eases the visualization of abnormal cells within organs. Some researchers are testing quantum dots to track individual glycine recep-tors in living neuronal cells in parts of the brain that are difficult to reach.[8] The use of quantum dots could help in the development of better drugs for a range of diseases such as depression and schizophrenia. Another potential application is treatment of breast cancer.[9] It may take 5 to 10 years before quantum dots find use as markers on antibodies for diagnosis. Due to their small size, it remains unclear whether quantum dots can also randomly succeed in penetrating healthy cells and cause damage to cellular structures such as DNAs.

Another type of nanostructure known as the magnetic nanoparticle may offer a unique alternative to chemotherapy and radiotherapy to cure certain forms of cancer. The basic idea is to use coated magnetic nanoparticles such as iron oxide that are selectively absorbed by tumor cells but ignored by most healthy cells. As a result, magnetic nanoparticles find themselves trapped in tumor cells and oscillate under the application of a magnetic field from outside the body. The repeated oscillation generates enough heat to destroy the tumor cells. This method called *magnetic fluid hyperthermia* has been tested as a cure of an aggressive form of brain cancer and is presently in clinical trials.[10]

Fullerenes or "buckyballs" are other types of nanoparticles that hold promise in biomedical applications. Fullerenes may succeed where many conventional drugs have failed: in crossing the blood–brain barrier to carry drugs from the blood stream into the brain.[11] Fullerenes may turn out to be unrivaled candidates in the fight against brain disorders such as Alzheimer's and Lou Gehrig's diseases. The efficacy of fullerenes or chemically modified fullerenes as drugs has not been fully estab-lished but the possibility that fullerenes may reach pharmacy shelves may not be that remote. The potential health side effects of fullerenes have been studied but not yet fully established.

D. Cosmetics

Nanoparticles made of zinc oxide (ZnO) are replacing conventional organic ultraviolet absorbers in some sunscreen lotions.* The advantage of using nanopar-ticles is that they do not scatter the visible light and this prevents skin whitening upon topical application of a sunscreen containing them. Some observers warn about the potential transmission of these nanoparticles through the skin and into the body, but their effects on human health remain to be identified.

III. TOXICITIES OF NANOMATERIALS

Generally speaking, toxicity issues related to nanomaterials are linked to multiple factors including chemical composition, size, shape, and surface chemistry. The most common paths for entrance of materials into the human body are inhalation through the respiratory tract, ingestion, injection into the blood stream, or transportation via the skin.

---

* Examples of sunscreen lotions include Wet Dreams, Wild Child, and Bare Zone.

The long-term safety issues and related risks, benefits, and costs of nanomaterials, drugs and devices have only started to be investigated. Therefore, one can only guess now about the risks that may be caused by using devices or drugs based on nano-materials. Concerns about the effects of some nanomaterials on vital organs and tissues have been discussed by scientists at international conferences and reported by the print media, but as of late 2003, very few studies have been published in peer reviewed medical and scientific journals.

From an historical perspective, research on the impact of particles on human health, particularly on respiratory effects, finds its roots in industrial manufacturing and processes such as gas exhaust from vehicles, coal, asbestos, man-made mineral fibers such as fiberglass, and ambient particulate matter in the atmosphere. This section, however, focuses on the toxicity of nanomaterials designed for biomedical purposes. It does not cover studies of ambient particulate matter resulting from gas exhaust or industrial activities.

Few studies have reported data on the toxicity of nanomaterials and most have not yet been independently replicated by other research groups. This means that most toxicity data are inconclusive. However, as funding for this type of research increases and new studies appear, it is expected that knowledge about the safety and toxicity of nanomaterials will evolve quickly.[12] As of fall 2003, the only peer reviewed studies of the health effects of nanomaterials concerned zinc oxide and titanium dioxide nanoparticles, fullerenes, and carbon nanotubes. The following section presents a brief overview of those studies.

A. Nanoparticles

In the scope of this section, nanoparticles are 100 nm in diameter or smaller. They can be produced by a number of methods: wet chemical processes (reacting chemicals in solution), mechanical processes (grinding and milling techniques), vacuum deposition, and gas phase synthesis. Depending on the method of fabrication, nanoparticles can be produced in a variety of sizes, chemical compositions, shapes, and with or without surface coating. Each of these factors influences how nanoma-terials interact with cells and tissues.

The available toxicity data on nanoparticles relate to ZnO and TiO_2 used in cosmetics. Their size is typically smaller than 50 nm and they act as ultraviolet absorbers and prevent skin whitening. It is important to note that ZnO and TiO_2 nanoparticles in sunscreens are often coated with other materials such as silicones, fatty acids, or zirconium to facilitate dispersion and to avoid the formation of clusters. In the presence of such coatings, cells and tissues are exposed primarily to the organic outer molecules rather than the inner cores made of ZnO or TiO_2. The presence and nature of coatings (which are not easily identifiable in the formulations of commercial products due to the need to retain trade secrets) may affect how the nanoparticles react with the skin. Questions have recently arisen whether the small size of ZnO or TiO_2 nanoparticles used in sunscreen lotions will allow them to accidentally penetrate the skin to an extent that can damage cells and eventually DNA.

A 1997 *in vitro* study revealed that under certain conditions both TiO_2 and ZnO nanoparticles can catalyze damage to DNA, although the fate of the nanomaterials through the skin was uncertain.[13] A more recent study on the distribution of sunscreens on skin showed that ZnO and TiO_2 nanoparticles cannot be detected in the human epidermis and dermis and remain on the outermost surface of the skin. This suggests that these nanoparticles do not travel through the skin.[14] The limited number of available studies prevents reaching a definite answer on the health effects of TiO_2 and ZnO nanoparticles.

B. Fullerenes

Fullerene molecules are 1 nm in diameter and similar in shape to footballs. Initial studies conducted by the University of Arizona and the Arizona Cancer Center in 1993 to determine the carcinogenic effects of uncoated fullerenes showed that at a dose level of 200 µg — thought to be a likely human exposure — fullerenes do not cause benign or malignant tumors on mouse skin even after repeated administration for a 6-week period.[15] A study from the Japanese National Institute of Health Sciences examined the effects of fullerenes on mouse embryos.[16] Toxic oxygen species produced by fullerenes induced cell damage to embryos at a dose of 50 mg/kg.

Uncoated fullerenes are poorly soluble in water. To be attractive for pharmaceutical applications, fullerenes are generally coated with a broad variety of organic molecules that increase their solubility in water and body fluids. No comprehensive data on the effects of coated fullerenes on cells and tissues are yet available.

C. Carbon Nanotubes

Skin exposure to carbon materials is known to increase the incidence of skin diseases such as carbon fiber dermatitis, respiratory tract infections, chronic bronchitis, pneumonia, and eventually cancer. Workers who are repeatedly exposed to high levels of carbon materials are most at risk. In the light of those findings, researchers have started to ask whether this carbon–skin disease relationship applies also to carbon nanotubes. Initial dermatological testing by the University of Warsaw to determine the effects of exposure of human skin to single-wall carbon nanotubes showed that the nanotubes do not cause skin irritations or allergic reactions.[17]

But, a recent joint study by West Virginia University, the National Aeronautics and Space Administration (NASA), and the National Institute of Occupational Safety and Health (NIOSH) found signs of toxicity after the exposure of single-wall carbon nanotubes to human cells *in vitro*.[18] They concluded that the toxicity response to human skin cells was due to the presence of iron, a by-product of nanotube fabrication, rather than carbon nanotubes per se. It is well established that iron loading in cells is a risk factor for certain cancers and infectious and inflammatory diseases of the skin, liver, and heart.

Studies have also been conducted to demonstrate the pulmonary toxicity of carbon nanotubes. Preliminary studies from the University of Warsaw published in 2001 on the health effects of carbon nanotubes followed the same procedure used to investigate asbestos-induced diseases. Carbon nanotubes did not exhibit abnor-malities or inflammation in guinea pig lungs.[19] These results have been challenged by more recent studies on the pulmonary toxicity of single-wall nanotubes in mice. A joint study conducted by NASA's Johnson Space Center and the University of Texas Medical School reported that single-wall carbon nanotubes (0.5 mg) injected directly into the lungs of mice led to the formation of microscopic nodules after a week.[20] These small nodules — which may eventually cause more serious lesions — persisted and were more pronounced after 3 months. A toxicology research team at DuPont conducted similar experiments independently.[21] Instead of injecting car-bon nanotubes directly into the lung, they placed them in the trachea — the tube that connects the throat to the lung. They found that after high doses of carbon nanotubes — equivalent to 5 mg/kg of weight — 15% of the rats died. The DuPont researchers suggested that death was caused by suffocation since the nanotubes tended to clump together and block the respiratory path. They also observed that exposure to single-wall carbon nanotubes led to cell injuries in lungs due to the formation of nodules. However, the nodules were not persistent beyond a month after instillation. This led the team to conclude that the nodules were reactions to foreign substances (the injected carbon nanotubes) and not necessarily the results of toxic reactions.

These studies are preliminary and present limitations, some of which are cur-rently under investigation. The experiments consisted of injecting carbon nanotubes into the lung or the trachea. However, in actual use, carbon nanotubes would most likely reach the lungs only if they were inhaled in the form of airborne particles. Initial studies on the handling of unrefined single-wall carbon nanotubes suggest that the aerosol exposure level in the laboratory is low.[22] Two key questions remain unanswered. What is the acceptable level of exposure before lung damage appears in humans? At what duration and frequency of exposure to carbon nanotubes can lung damage be detected?

Risks associated with nanomaterials will be fully characterized when both haz-ards and exposure levels are determined. A hazard becomes a risk when organisms are exposed to significant doses of nanomaterials at minimum frequencies. Con-versely, a hazard does not become a risk when the level of exposure is low and the frequency of exposure is rare. Frequent exposure at low levels can possibly present some risks due to accumulation effects over time. Exposure levels to nanomaterials can vary, depending on the original form of the exposed material and the method of exposure — breathing, direct dermal exposure, or injection.

As discussed earlier, initial identification of nanomaterial hazards has been accomplished by dose–response studies that observe where nanomaterials propagate in the body from a site of entry (the respiratory or digestive tract, via injection into the blood stream, or through direct dermal contact) to remote organs. This type of study provides information on how much nanomaterial the body uptakes and, in the case of bioaccumulation, which tissues react and how they do so in the presence of nanomaterials. An important issue is defining acceptable exposure levels through *in vitro* and *in vivo* exposure–response studies that provide data on chronic and acute exposure to a given nanomaterial. Both dose–response and exposure–response stud-ies are necessary to assess risks. Since only a limited number of such studies have been performed on nanomaterials, the characterization of the human health risks of the various types of nanomaterials is far from complete.

IV. STAKEHOLDERS' POSITIONS ON SAFETY AND REGULATION OF NANOMATERIALS

Stakeholders can be broadly defined as persons or groups of persons (or even animals and plants) affected by or able to affect the risks inherent in new technologies. Because biomedical nanotechnology and its applications are still in infancy, the precise identifications of all stakeholders and their positions relative to safety and regulation remain for the most part blurred. Most of the disagreements among the different stakeholders reported to date in the media have centered primarily on the *fear of risk* — perceived risks rather than technically understood risks — since the commercial production of nanomaterials is nascent and the hazards are not yet fully understood.

Complicating this situation is the fact that stakeholders differ in their natural tendencies to be aware of, organize around, or publicize biomedical nanotechnology risks. To some stakeholders, the amplification of the fear of risk ensures public exposure and may be valuable. For these stakeholders, more is better. For the mass media, bad news is good news because it generates greater sales. Certain nongovernmental organizations such as consumer groups and environmentalists use the fear of technological hazards as a driving point to collect and retain members.

To other stakeholders, the fear of risk is worrisome and less is better. For example, government agencies that promote advances in science and technology through research grants and public outreach activities, large corporations and start-up companies whose advanced products constitute their competitive advantages, and university scientists who heavily rely on public funds for conducting their research are potentially vulnerable to public fears and media reports about the hazards of nanomaterials.

This section presents an overview on how the different stakeholders (the scientific community, industry, citizen interest groups, the public, and governmental agencies) have begun to address issues linked to the safety and regulation of nanomaterials. For each stakeholder category, we focus upon three important aspects: (1) how the stakeholder is affected by the potential risks and benefits in engaging in activities linked to nanomaterials; (2) how the stakeholder may influence or affect the risks; and (3) the types of activities in which the stakeholder has engaged to date regarding nanomaterials safety and regulation.

A.   Scientific Community

The scientific community includes scientists and engineers who are active in understanding basic phenomena at the nanoscale, designing and fabricating new nanomaterials and devices, and developing applications for medical purposes. The scientific community has greatly benefited from funding support provided by governments and made significant breakthrough discoveries in medical applications of nanomaterials. For example, the U.S. federal budget for nanotechnology research and development increased from $116 million in 1997 to $862 million in 2003.[23]

Equivalent trends can be observed in Europe where public investment in nanotechnology rose from $126 million in 1997 to $650 million in 2003, and in Japan where the total increased from $120 million in 1997 to $800 million in 2003.

If public fears about nanotechnology develop because of uncertainty regarding the potential risks, it is likely that public pressure will slow down the increases of funds for research or, worse, may lead to funding cuts or even the elimination of funds for certain activities.

Scientists play a special and influential role in the debate about the risks because of their expertise in the development of nanomaterials and in the assessment of attendant risks. As discussed earlier, scientists have started to report data on the effects of nanomaterials on cells and tissues at international conferences and are publishing results of preliminary studies in peer reviewed journals. These studies have occasionally been covered by the mass media such as the *New York Times* and the *Washington Post*, making nanomaterial risk a broader public issue by heightening awareness of business managers, politicians, and the public at large. Scientists have also been called to testify before congressional committees that focus on the potential impacts of nanotechnology on societal, ethical, and environmental issues, thus influencing future policies and research funding.

The Achilles' heel of the scientific community is that it is more fragmented than some of the other stakeholder groups. Most of the funding in science is focused on narrowly defined areas. Typically, as long as funding continues, scientists will probably not act collectively on broad issues of toxicology that may only exert indirect impacts on their work and may affect their funding. Generally speaking, the scientific community tends to be less organized politically than citizen or industry groups.

Nevertheless, scientists have begun to study the health risks associated with inhalation and dermal exposure to nanoscale materials. Toxicologists who examined the health impacts of asbestos, quartz particles, and fume exhausts have started to investigate nanomaterials.[24-33] The scientific community has acknowledged the paucity of available data and the need for further investigation of the reactivity of nanomaterials with living organisms, the human body, and the environment; it favors the collection of additional and comprehensive data regarding the risks associated with the production and use of nanomaterials. Discussions of risk assessment and policy are more frequently added to the agendas of national and international scientific conferences.

B. Industry

Politicians often tout nanotechnology as a driving force of the new economy. However, in comparison to other industries, nanotechnology is still a tiny (but growing) field. Nanotechnology venture capital funding in the U.S. was estimated to have increased from less than $10 million in 1997 to $300 million in 2003.[34] Corresponding increases in the number of nanotechnology-related patents and formation of companies involved in the production and commercialization of nanomaterial products have occurred. The integration of the *nano* prefix into company names and products can be viewed as a winning marketing strategy to attract funding and increase sales.

However, all of this progress is still quite frail. Industry could pay a high cost if the use of medical devices or drugs derived from nanotechnology presents risks, causes harm, or creates serious public health or environmental problems. Stringent regulation of the production, labeling, and use of nanotechnology-based products, and liability costs are likely to affect business growth and investors' perceptions of the nanotechnology market.

Industry is well positioned to contribute to the debate about safety and regulation of nanomaterials. First, the companies in the industries most connected to nanomaterials for medicine are well funded and politically active. For example, according to the nonpartisan Center for Responsive Politics, the pharmaceutical industry spent about $30 million for individual contributions, political action committee contributions, and soft money contributions to U.S. campaigns in the 2001–2002 election cycle.[35] In the same period, chemical manufacturers contributed another $7 million and the health industry as a whole spent a whopping $95 million.[36] These companies also operate in many congressional districts, giving them preferred access to many congressional members. Finally, these companies have scientific experts who are capable of preparing reports and providing testimony on nanomaterial risk.

To date, the nanotechnology industry is monitoring the stakeholder activism about risk and initiating discussions with governmental bodies about the impacts of nanomaterials on human health and the environment. Efforts to identify potential risks from the industry perspective are present but isolated. Some large corporations such as DuPont are presently investigating the health impacts of some types of nanomaterials. In the summer of 2003, the Nanobusiness Alliance, a U.S.-based trade group of nanotechnology companies, initiated its Health and Environmental Task Force.[37,38] Composed of scientists, government staffers, business leaders, venture capitalists, and lawyers, the group aims to develop standards and best practices for the production and disposal of nanomaterials. No results of their investigation have been released to date.

Insurance companies are also interested in mapping the nanomaterials' risk landscape.[39] They will likely seek to benchmark figures regarding potential damage caused by nanomaterials, and their impacts on workers, patients, children and the elderly, and on wildlife. Insurance companies acknowledge the need for a more systematic and complete risk assessment of nanomaterials, for clarification regarding regulation or guidelines for the production and use of nanomaterials in commercial products, for approval, certification and labeling requirement of new products, and for national and international standardization. By doing so, insurance companies wish to identify the types of industry sectors and countries that are most likely to be affected by risks and liability issues.

C. Citizen Interest Groups

Citizen interest groups such as those focused on environmental or public safety concerns are typically organized to represent those affected by some activity in those areas. Generally, such groups respond to activities that they see affecting their rights and personal lives. For example, a public safety group may feel that the pharmaceutical industry's rapid expansion intended to produce nanomaterials to deliver new medicines violates the public's right to safety and should be slowed or stopped until credible evidence of safety is presented.

Citizen interest groups also may advocate the status quo until clear benefits of new technologies are proven. Such groups played key roles in limiting the expansion of genetically modified crops for use in human foods because the consumer benefits such as better and less expensive nutrition were not perceived to outweigh the potential risks to health and the environment. Sometimes, however, citizen interest groups push to expand rights, for example, a consumer group may advocate giving access to a new nanotechnology-based therapy to individuals with serious health needs such as AIDS patients, thereby extending access to a broader public instead of limiting it exclusively those who can afford it.

Certain citizen interest groups are beginning to express opposition to nano-technology. A number of factors have triggered this movement. First, the lack of definite scientific consensus on the safety of nanomaterials raises questions. Another factor is an emerging distrust that governments will ensure the safety of nano-technology because the U.S. and EU have not yet delineated clear rules about nanomaterial safety. All of these issues contribute to create confusion and heighten opposition among "nanoskeptics," allowing interest groups to take their opposition into the public sphere.

Public personalities such as Prince Charles[40,41] and Caroline Lucas[42,43] of the Green Party in the U.K. recently led highly publicly visible campaigns against nanotechnology and called for caution. The Canadian-based ETC (erosion, technology, and concentration) group called for a moratorium on the production of nano-materials.[44,45] Greenpeace[46] and GeneWatch U.K.[47] also expressed concerns regarding the potential hazards of nanotechnology. They encourage more inclusive debates among scientists, government officials, and the public to define policies.

The interest of citizen groups in nanotechnology is still somewhat limited, but their voices are heard by scientists, policy makers, and journalists. Despite the publicity, the nanotechnology debate still remains largely unknown to the silent majority — the public.

D. The Public

The public is a key stakeholder in the sense that as consumers and users of nanotechnology applications, they are ultimately affected by the risks and the benefits. Biomedical applications of nanotechnology promise to improve life styles and lead to better medical treatments, particularly for diseases for which existing treatments have undesirable side effects or diseases that have no treatments at present.

Since the public is for the most part unaware of nanotechnology, one can only guess how the public will receive and perceive consumer products derived from nanotechnology. Few "nano" products exist to date and the few that are available have not encountered significant public resistance. However, as applications grow and as other stakeholders become active, consumers may develop stronger acceptance or resistance to nanotechnology-enhanced products.

The genetically modified food situation in Europe, particularly in the U.K., is a case in point. Public fears can determine the progress or lack of progress of technological applications. Throughout the 1990s, the public generally was oblivious to the existence — and risks — of genetically modified foods until the publication of a controversial, nonrefereed study reporting that rats fed with genetically modified potatoes suffered damage to vital organs led to multiple news stories, a general public backlash, and a swift governmental response. In 1998, the European Commission placed a moratorium on the importation and cultivation of genetically modified foods by member states. Public fears were inflamed more by accusations than by scientific knowledge, but the outcome was clear: genetically modified foods were driven out of the market. The moratorium was partially lifted in July 2003, but strict labeling requirements were placed on all foods derived from genetic modifications.

The point is that while the vast majority of the public may typically remain silent on a number of issues; at times it weighs in to significantly affect the pace of technological development. Sometimes it does so through consumption decisions, such as in the case of the rejection of genetically modified foods in Europe. At other times, the public enters the political arena through referendum or demonstration, although these are not common approaches. To date, the public has not been a major player in the debate about the risks of nanotechnology.

E. Government Funding and Regulatory Agencies

Governments play dual roles in the development of nanomaterials and the assessment of their risks. Governmental agencies assume great risks risk in the development of nanotechnology by underwriting significant amounts of research through large-scale grants to scientists and scientific institutions. In this way, governments also indirectly promote the development of industry and accelerate the transfer of fundamental knowledge of nanoscale science and technology to marketable applications.

A government also plays the role of protector of the public from dangerous situations. With nanotechnology, particularly its production and applications in drugs and other products, the government plays a watchdog role to ensure that the overall risks to production employees and users are not inordinate. The government's influential power comes from its ability to regulate and even to disallow the development and use of nanomaterials. Several governments have launched research programs to assess the technical risks of such materials.

1. U.S. Government Initiatives

In August 2003, the National Nanotechnology Coordination Office and the Office of Science and Technology Policy convened the formation of the Interagency Working Group on Nanotechnology Environment and Health Implications (NEHI) whose role is to investigate how current regulatory paths can cover the production and use of nanomaterials, including workplace regulation and environmental and health risks.

Earlier that year, the U.S. Environmental Protection Agency (EPA) launched a call to academic and not-for-profit organizations for proposals concerning the impacts of manufactured nanomaterials on human health and the environment.[48] The total anticipated funding is $4 million and the initiative focuses on the studies of the toxicity of manufactured materials, their environmental and biological transport, their exposure, and their bioavailability.

The National Toxicology Program of the National Institute of Environmental Health Science is investigating nanoscale materials for toxicological studies.[49] The National Toxicology Program was established in 1978 by the U.S. Department of Health and Human Services. Under this program, the toxicological studies focus on semiconductor nanocrystals such as quantum dots, carbon nanomaterials such as fullerenes and carbon nanotubes, and metal oxide nanoparticles such as TiO_2. The program has the task of evaluating the health impacts of environmental and occupational exposures to chemicals and various physical agents. For example, it generates and collects tests on chemicals that may be related to health problems such as cancer, genetic and reproductive toxicity, birth defects, and brain and nervous disorders.

The National Toxicology Program reports to federal regulatory agencies such as the Food and Drug Administration, the National Institute for Occupational Safety and Health, and the Agency for Toxic Substances and Disease Registry. The program serves also as a source of information for the EPA and the Consumer Product Safety Commission and the information it gathers may be used for recommendations for future regulations of nanoscale materials.

The Center for Biological and Environmental Nanotechnology (CBEN) based at Rice University in Texas and funded by the National Science Foundation has started to investigate the impacts of nanomaterials on biological systems and on the environment.[50] Results of these studies are expected to be released soon. CBEN has also engaged discussions on the broader societal implications of nanotechnology through its annual workshops that convene scientists, engineers, social scientists, venture capitalists, lawyers, and advocacy groups. In 2003, Professor Vicki L. Colvin, CBEN's director, testified before the U.S. House of Representatives' Science Committee on the social, ethical, and environmental issues of nanotechnology.[51]

2. Government Initiatives in Europe

In June 2003, the Royal Society and the Royal Academy of Engineering, the most prestigious scientific institutions in the U.K., created a working group on nanoscience and nanotechnology commissioned by the government's Office of Science and Technology.[52] The group's goal is to determine the need for new regulations regarding the control of nanotechnology, specifically in the areas of health, safety, toxicity of nanoparticles, and ethics. The study is meant to engage various stakeholders: academia, industry, special interest groups, and the public. The public will be consulted through online discussions, focus group consultations, and surveys. The final report of this independent study was released in the summer of 2004 and is available at www.nanotec.org.uk/final report.htm.

The European Commission initiated a series of risk assessment studies[53–55] designated the Nano-Pathology Project, the Nanoderm Project, and the Nanosafe Project in 2002. The Nano-Pathology Project will define the roles of microparticles and nanoparticles in biomaterials-induced pathologies. *In vitro* studies of the effects of nanoparticles on cell structure and function, *in vivo* studies to simulate exposure to nanoparticles, and clinical studies are presently in progress. The Nanoderm Project investigates the fate of TiO_2 and other nanoparticles used in body care and household products. Issues such as uptake and clearance of nanoparticles and their reactivity with cells and tissues are under investigation. The Nanosafe Project aims at determining the best ways to handle risks involved in the production, handling, and use of airborne nanoparticles in industrial processes and consumer products. After completion of the project tasks, recommendations will be offered for regulatory measures and codes of practice in the workplaces to limit the potential adverse effects of nanoparticles on workers.

Most of the governmental initiatives in the U.S. and in Europe were created recently, and results are not yet publicly available. Some data are expected to be released in 2005 and will constitute the basis for assessing the risks and benefits of certain applications of nanotechnology and for drafting pertinent regulations.

V. POTENTIAL PATHS FOR REGULATION

A parallel issue to risk assessment of nanotechnology is the pending question of regulation. Through 2004 no endorsed regulatory policy regarding nanomaterials exists. Stakeholders' opinions regarding nanotechnology regulation — or whether there is a need for it — are divergent. Generally speaking, governments have noted the need for a legal framework to address the ethical and social consequences of other emerging technologies such as genetic engineering on public safety and impacts on workers and the environment, privacy, and security. Governments tend to be reactive about such matters and consequently regulate in the aftermath of a dramatic accident or other unintended incident related to use of a new technology. Quite unique to nanotechnology is the fact that government leaders and decision makers in the U.S. and Europe have commented on the importance of being proactive and addressing the social and ethical issues of nanotechnology in parallel with its development.

The following section presents overviews of the different regulatory paths for nanotechnology, from least to most stringent: (1) regulation through the market; (2) application of current regulations for related products such those applying to drugs, cosmetics, chemicals, the environment, and the workplace; (3) regulation through accidents; (4) regulatory capture; (5) self-regulation; and (6) bans.

A. Regulation through the Market

Regulation through the market means that innovation progress is left to market forces. In such scenarios, innovation occurs only if consumers see economic benefits to buying technology-enhanced products. For example, if consumers feel they can realize benefits from digitally enhanced photography by easily transmitting images to remote parties, they will purchase digital cameras and related equipment. Conversely, if consumers do not see the benefit of switching to high-tech products, they do not purchase the products and they eventually disappear. In this manner, the purchasing behavior of consumers drives certain technologies to advance and others to perish.

Safety is also handled in a similar fashion. If consumers desire a certain level of safety, they will pay for it; as a result, safety is thrust onto products. If consumers perceive front air bags to be important safety features of an automobile, they will demand the bags and manufacturers will in turn be forced to offer such safety devices. Of course, if consumers do not demand a safety feature, it will not appear in the market. As with product innovation, safety is a function of consumer demand. The government's role is primarily limited to overseeing the functioning of the market — protecting property rights and preventing deceptive practices such as false advertising. Occasionally, the government's role may expand to force companies to disclose certain information such as the accuracy of accounting statements or safety data related to products.

Unfortunately, several deficiencies are involved in allowing consumers to dictate levels of safety. First, consumers may not know about the safety aspects of a given technology. For example, while some cosmetic companies have added nanoparticles to their sunscreen lotions, most consumers are likely to be uninformed about their use and potential impact on human health. This means many consumers make purchase decisions without adequate information. Second, even if consumers know about a technology, they may be unable to completely understand its safety effects. As noted earlier, even scientific experts have not reached definite conclusions about the toxicity of nanomaterials. Based on the current state of knowledge, it is highly unlikely that consumers will be able to make informed decisions about the risks of nanoproducts.

When products have latent risks, for example, slight and cumulative effects that arise from exposure to pesticides, consumers often underestimate the long-term effects even when they are shown to be injurious. Finally, the consumer-based model ignores nonconsumers who may be negatively affected by product purchases. A classic example is second-hand smoke from tobacco users. Nonusers play no part in the market transaction but nonetheless suffer the effects of the consumer's purchase and use of the product. Economists call this a *market failure* — in this case it is a negative externality — and the general remedy is some type of government regulation.

B. Application of Current Regulations

Nanomaterials such as nanoparticles, quantum dots, nanotubes, and others can be viewed technically as chemicals. At present, more than 20 million known chemicals are indexed by the American Chemical Society's *Chemical Abstract Service*.[56] Of those, roughly 6 million are commercially available. Only about 225,000 are inventoried and regulated. Regulatory agencies generally attempt to focus on chemicals such as benzene, lead, and mercury, that have the potential to cause harm. Most chemicals are unregulated.

Only very recently have governments considered regulation of nanomaterials. An umbrella of mechanisms is already in place for assessing and regulating the hazards new materials impose on human health and the environment. The crux of the issue is to determine whether existing regulatory mechanisms are adequate to regulate new nanomaterials and devices or whether they require amendments.

In the current regulatory framework in the U.S., nanomaterials-based substances incorporated into consumer products would be regulated under the Federal Hazardous Substance Act administered by the US Consumer Product Safety Commission[57] and no premarket certification or approval is required. However, the use of the new substance would be controlled according to risk of exposure.

Discussions about regulation have started in the U.S. During a workshop in the fall of 2003, the Woodrow Wilson International Center for Scholars based in Washington, D.C. gathered experts in public policy and science and engineering to discuss whether the Toxic Substance Control Act (TSCA, administrated by the EPA), an existing framework that has regulated toxic substances in the U.S. since 1976, would apply to nanotechnology.[58] More specifically, the participants considered, among other things, whether the TSCA would apply to the safety of and exposure to nanomaterials such as nanoparticles, fullerenes, and carbon nanotubes. If the "Significant New Use Rule" of the act applies, the EPA could investigate the effects of nanomaterials prior to their manufacture and require postproduction testing for exposure. Manufacturers, processors, and importers would be subject to regulation. Conclusions published in the report[59] titled *Nanotechnology and Regulation: A Case Study Using the Toxic Substance Control Act* stated:

> The very nature of nanotechnology … is likely to challenge the existing regulatory structure and cause confusion both on the side of industry and the government concerning the role of regulation …. A wrong or ill-conceived approach to regulation could have enormous economic consequences ….

Typically, the U.S. Food & Drug Administration (FDA) and the Department of Agriculture (USDA) regulate foods and food packaging. Drugs, food additives, pharmaceuticals, and diagnostic and therapeutic devices are regulated by the FDA. Drugs, food additives, and food coloring require premarket approval from the FDA. The lack of well defined nomenclature for identifying nanomaterials makes regulation using the current acts tricky.

In conclusion, it remains to be seen whether the current regulations can apply to the production and use of nanomaterials. In some cases, it may be appropriate to revise existing legislation, classifications, and labeling standards and to make new recommendations regarding the manufacture, use, and disposal of nanomaterials and their impacts on human health and the environment.

C. Regulation through Accident

Many feel that regulation through (resulting from) accident[60] has been the *modus operandi* behind numerous public policies. It means that incentives for new safety regulations are triggered by problems that were not anticipated. Such a policy may have some rational basis. When the expected benefits and costs of emerging technologies are very difficult to estimate, as is the case for nanotechnology, the conventional public policy framework of benefit–cost analysis is likely to produce inaccurate results. That is, the option chosen from benefit–cost analysis is equally as likely to be good or bad for society. In such cases, it may be best to scrap the formal analysis altogether and wait for incidents to occur. As a result, regulation and legislation are reactive and new safety measures are established only after accidents happen. Unfortunately, accidents may occur too late to prevent irreversible impacts.

Regulation through accident is being challenged at many levels, particularly by environmentalists and citizen groups, because it tends to erode trust among the different stakeholders. *Post hoc* regulation without *a priori* prevention can injure many people, nonhumans, and the natural environment. This is reckless behavior by the purveyors of the technologies, especially when the probable effects are fatal. Furthermore, when the harm such as that caused by exposure to chemicals is long-term, many victims may be exposed before detection is achieved.

Accidents tend to attract much media exposure, but the exposure inevitably paints negative images of technologies. For example, the media reported the incidents at Bhopal and Chernobyl instead of publishing balanced reports about the relative risks and rewards of use. Regulation by accident often leads to a confusing locus of responsibility, complicating future rectification. A good example is the nuclear power plant accident at Three Mile Island in Pennsylvania, the blame for which was spread widely among government agencies, the operator, the company that built the reactor, and others.[61] If the reports were accurate, the best entity to implement a remedy is unclear. Should the federal government have provided more oversight? Should the utility company have hired more qualified people? Should the designer of the reactor have created a different design? Finally, the solutions that emerge from *post hoc* regulation are often only politically expeditious when proposed and not over the long term.

After a technology-related accident, especially a serious one, regulators, politicians, business managers, and others often scramble to "do something" while the media shine bright lights on them. "Doing something" may entail a regulatory solution that can be passed and allow regulators to quickly say they "did something," but this type of solution does not offer an effective long-term solution.

What may be more worrying, particularly in North America, is the fact that this *post hoc* regulatory approach feeds into a seemingly flawed litigation system. While the intent of litigation is to give individuals the tools to enforce their rights to be protected against accidents through monetary compensation, it can become excessive. In September 2003, the Manhattan Institute for Policy Research, a New York think tank, published a report titled *Trial Lawyers Inc.: A Report on the Lawsuit Industry in America 2003.*[62] That revealed the astonishingly high costs of litigation for businesses and the ensuing revenues received by trial lawyers. For example, settlements for tort litigation in the U.S. exceeded $200 billion in 2001, of which $39 billion went to trial law firms. The magnitude of litigation costs has to an extent created a new form of industry, hence the addition of "Trial Lawyers Inc." to the title of the report. Law firms handling litigation involving medical technology received about $1.4 billion for asbestos cases and another $1 billion for medical malpractice lawsuits in 2002. The report also notes that "for the lawsuit industry as a whole, less than half of all dollars actually go to plaintiffs, and less than a quarter of dollars actually go to compensate plaintiffs' economic damages."

If cautiously used, litigation procedures can secure consumers' rights and enhance business performance by providing incentives to be prudent in using technologies. However, experience shows that regulation through litigation provides meager rewards to plaintiffs and becomes excessively costly to industry. In summary, abuse of the litigation system is economically and socially counterproductive because it tends to delay the introduction of new products and increases product costs. In the light of such evidence, there are good reasons to question whether regulation through accident or litigation is the most effective and efficient tool to regulate or make amends to injured parties.

D. Regulatory Capture

Regulatory capture is a phrase coined by George Stigler, an economist and Nobel Laureate in 1971 to describe a situation when a company seeks regulation instead of resisting it.[63] The logic is that regulations create barriers to the entry of new companies into an industry and the barriers enhance incumbent company profits. For example, a regulation that requires chemical companies to invest in a certain, perhaps very expensive, type of filter system to capture fumes creates a mandatory cost barrier that certain companies are better able to bear than others. In this way, the regulation creates differences among firms; certain firms can better compete against incumbents and entrants. This scenario is not typical across industries but is a possibility in the nanotechnology area.

In certain cases, companies will ask the government to regulate them in an effort to "manage" competition. An example is the practice of the U.S. Department of Defense to enter into contracts with a limited number of U.S. companies. Likewise, companies often engage in certain practices to make the regulators more responsive, for example, by providing detailed reports of research and other company information and hiring former regulators. It should be noted that while regulatory capture focuses on the private benefits of regulation, it does not necessarily produce regulatory outcomes in the public interest.

While regulatory capture historically has been most prominent in the defense and transportation industries, it also has occurred in some high technology industries. Most major pharmaceutical companies maintain close relationships with FDA regulators who approve new drugs. FDA approvals and other regulatory schemes such as patents act as mechanisms to fight against competitors — generic drugs in the case of major pharmaceutical companies. As nanotechnology pushes toward commercialization, it is possible that a number of companies will pursue regulations to lock out potential competitors.

E. Self-Regulation

On July 26, 1974, a group of eminent scientists published a letter in *Science* asking their peers working in the emerging area of recombinant DNA to join them in agreeing not to initiate experiments until attempts were made to evaluate the hazards. The scientists feared that recombinant DNA molecules could prove biologically hazardous.[64] Recombinant DNA, a discovery that marked the birth of genetic technology and biotechnology, involves joining parts of DNAs from different biological sources — viruses, bacteria, and animals — to produce hybrid molecules that can, for example, penetrate into bacteria and replicate. In the early 1970s, scientists felt that the biological properties of such hybrids could not be readily predicted.

A year later, that letter triggered an event that set a precedent in the history of science: a call for a temporary voluntary moratorium from scientists to stop research on recombinant DNA until they evaluated the potential risks on human health and the Earth's ecosystems and defined guidelines for research to proceed. This voluntary moratorium directed to the scientific community served as a form of self-regulation. It was widely accepted and observed by the scientific community and lasted about a year. The moratorium was relaxed when safe working practices and regulations were put in place by 1976. In retrospect, most agree that these restrictions did not hamper the biotechnology industry boom in the early 1980s.

Two decades later, another technological breakthrough led to a somewhat similar scenario. In 1997, a few months after the birth in Scotland of Dolly, the first (and late) cloned sheep, the Federation of American Societies for Experimental Biology (FASEB), the largest coalition of biomedical scientists, the U.S., endorsed a voluntary 5-year moratorium on cloning human beings. In February 2003, FASEB approved an extension of the original voluntary moratorium for an additional 5 years.

Acknowledging that the pursuit of recombinant DNA research and cloning is a key step in understanding the fundamental processes of life such as deciphering the human genome and detecting diseases related to gene mutations, scientists tend to view self-regulation in the form of a temporary voluntary moratorium as an effective method of intervention to eliminate potentially harmful or unsafe procedures.

Business persons also favor self-regulation in certain scenarios. Self-regulation generally entails the norms and practices derived from leading companies or laboratories in a given field. These companies and laboratories sometimes desire to make their practices the norms because they have already achieved what they wanted. Self-regulation typically puts group pressure on companies and laboratories for compliance because if they are lax in their practices, government may step in with more stringent guidelines.

F. Ban

In 2003, a Canadian-based environmentalist pressure group known as ETC called for a ban on nanotechnology products. For fear of losing control over nanotechnology applications to human health and the environment, skeptics against nanotechnology progress asked that research and development in that area be stopped. This ban is quite different from the temporary voluntary moratorium mentioned earlier. A ban prohibits all activity by law; a temporary moratorium delays activities for an authorized period of time. In practical terms, a ban means that nanotechnology would cease to exist — no research, no production, no products.

Pressure groups invoke the precautionary principle to justify their calls for bans on nanotechnology products. The precautionary principle is a way of thinking about risks. In short, it means that activities are stopped in the face of uncertainty. Proponents of the precautionary principle welcome bans as measures that reflect public interest, for example, protecting the public from exposure to hazardous materials. Politicians often cite the precautionary principle as a "better safe than sorry" practice.

The precautionary principle contrasts with the more conventional and long-practiced way of assessing risks: learning by doing and drafting policies. While the precautionary approach and the more conventional method both strive to balance progress and caution, the precautionary principle is by nature more conservative. Although pressure groups generally invoke the precautionary principle to halt activities, use of the principal does not automatically lead to a ban.

The precautionary principle has become a topic of controversy because it has profound implications. For example, it puts the burden to prove nonharm on the proponent of an activity, whether industry or regulatory agency, rather than on potential victims. Economists criticize the precautionary principle for its lack of consideration of opportunity costs, that is, the value of the best alternative. For example, in assessing the costs of a ban on nanotechnology, one should consider what advances in medical technology might be foregone because of the cessation of the research.

Europe and the U.S. take different approaches to the precautionary principle. The Maastricht Treaty on the European Union states that the "community policy on the environment ... shall be based on the precautionary principle that preventive actions should be taken, that environmental damage should as a priority be rectified at source and that the polluter pay."[65] In 2000, the European Commission suggested that the precautionary principle be invoked where the "identification of potentially negative effects resulting from a phenomenon, product or procedure" or where "a scientific evaluation of the risk because of the insufficiency of the data, their inconclusive or impressive nature, makes it impossible to determine with sufficient certainty the risk in question."[66] The situation is different in the U.S. where the precautionary principle has not been formally expressed in legislation, although some argue that the spirit of the principle is present through the requirement for premarket approvals of new pharmaceuticals, foods, additives, pesticides, and chemicals.

Ultimately, the decision to use the precautionary principle has less to do with science than with politics. Under pressure from the media, nongovernmental organizations, and the public, governments may be forced to take the precautionary principle into consideration. Even if the EU and U.S. decide to ban nanotechnology products, progress will likely continue somewhere else; governments that decide to ban nanotechnology will probably fall far behind the technology frontier and scientific research and commercial development will experience negative impacts.

REFERENCES

1. TJ Webster, C Ergun, RH Doremus, RW Siegel, and R Bizios. Enhanced osteoclast-like cell functions on nanophase ceramics. *Biomaterials* 22: 1327–1333, 2001.

2. TJ Webster, RW Siegel, and R Bizios. Nanoceramic surface roughness enhances osteoblast and osteoclast functions for improved orthopedic/dental implant efficacy. *Scripta Mater* 44: 1639–1642, 2001.

3. J Lin-Liu. Chinese researcher ready to 'bring nano bones to the world'. *Smalltimes*, July 1, 2003. http://www.smalltimes.com/document_display.cfm?document_id=6300 &keyword=bone%20and%20implants&summary=1&startsum=1.

4. W Xiaohong, M Jianbiao, and F Qingling. Skeletal repair in rabbits with calcium phosphate cements incorporated phosphorylated chitin. *Biomaterials* 23: 4591–4600, 2002.

5. JD Harterink, E Beniash, and SI Stupp. Self-assembly and mineralization of peptide–amphiphile nanofibers. *Science* 294: 1684–1688, 2001.

6. DR Larson, WR Zipfel, RM Williams, SW Clark, MP Bruchez, FW Wise, and WW Webb. Water-soluble quantum dots for multiphoton fluorescence imaging *in vivo*. *Science* 300: 1434–1436, 2003.

7. M Bruchez Jr., M. Moronne, P Gin, S Weiss, and A Alivisatos. Semiconductor nanocrystals as fluorescent biological labels. *Science* 281: 2013–2016, 1998.

8. M Dahan, S Lévi, C Luccardini, P Rostaing, B Riveau, and A Triller. Diffusion dynamics of glycine receptors revealed by single-quantum dot tracking. *Science* 302: 442–445, 2003.

9. X Wu, H Liu, J Liu, KN Haley, JA Treadway, JP Larson, N Ge, F Peale, and MP Bruchez. Immunofluorescent labeling of cancer marker Her2 and other cellular targets with semiconductor quantum dots. *Nat Biotechnol* 21: 41–46, 2003.

10. K James. Nanoparticles fight cancer cells with lethally high 'fever'. *Smalltimes*. September 11, 2003. http://www.smalltimes.com/document_display.cfm?document_ id=6606&keyword=nanoparticles%20and%20cancer&summary=1&startsum=1.

11. C Stuart. Companies enter deals to test drugs for diseases in the brain. *Smalltimes*. December 3, 2003. http://www.smalltimes.com/document_display.cfm?document_id =7032.

12. VL Colvin. The potential environmental impact of engineered materials. *Nat Biotechnol* 21: 1166–1170, 2003.

13. R Dunford, A Salinaro, L Cai, N Serpone, S Horikoshi, H Hidaka, and J Knowland. Chemical oxidation and DNA damage catalysed by inorganic sunscreen ingredients. *FEBS Lett* 418: 87–90, 1997.

14. J Schulz, H Hohenberg, F Pflücker, E Gärtner, T Will, S Pfeiffer, R Wepf, V Wendel, H Gers-Barlag, and K-P Wittern. Distribution of sunscreen on skin. *Adv Drug Delivery Rev* 54 (Suppl. 1): S157–S163, 2002.

15. MA Nelson, FE Domann, GT Bowden, SB Hooser, Q Fernando, and DE Carter. Effect of acute and subchronic exposure of topically applied fullerene extracts on the mouse skin. *Toxicol Ind Health* 9: 623–630, 1993.

16. T Tsuchiya, I Oguri, YN Yamakoshi, and N Miyata. Novel harmful effects of [60] fullerene on mouse embryos *in vitro* and *in vivo*. *FEBS Lett* 393: 139–145, 1996.

17. A Huczko and H Lange. Carbon nanotubes: experimental evidence for a null risk of skin irritation and allergy. *Fullerene Sci Technol* 9: 247–250, 2001.

18. AA Shvedova, V Castranova, ER Kisin, D Schwegler-Berry, AR Murray, VZ Gandelsman, A Maynard, and P Baron. Exposure to carbon nanotube material: assessment of nanotube cytotoxicity using human keratinocyte cells. *J Toxicol Environ Health, Part A*, 66: 1909–1926, 2003.

19. A Huczko, H Lange, E Cako, H Grubek-Jaworska, and P Droszcz. Physiological testing of carbon nanotubes: are they asbestos-like? *Fullerene Sci Technol* 9: 251–254, 2001.

20. C-W Lam, JT James, R McCluskey, and RL Hunter. Pulmonary toxicity of single-wall carbon nanotubes in mice 7 and 90 days after intratracheal instillation. *Toxicol Sci* 77: 126–134, 2004.

21. DB Warheit, BR Laurence, KL Reed, DH Roach, GAM Reynolds, and TR Webb. Comparative pulmonary toxicity assessment of single-wall carbon nanotubes in rats. *Toxicol Sci* 77: 117–125, 2004.

22. AD Maynard, PA Baron, M Foley, AA Shvedova, ER Kisin, and V Castronova. Exposure to carbon nanotube material: aerosol release during the handling of unrefined single-walled carbon nanotube material. *J Toxicol Environ Health, Part A*, 67: 87–108, 2004.

23. U.S. National Nanotechnology Initiative. http://www.nano.gov.

24. DB Warheit. Nanoparticles: health impacts? *Mater Today* 7: 32–35, 2004.

25. DB Warheit and M Hartsky. Initiating the assessment process for inhaled particulate materials. *J Exposure Anal Environ Epidemiol* 7: 313–325, 1997.

26. G Oberdörster, Z Sharp, V Atudorei, A Elder, R Gelein, W Kreyling, and C Cox. Translocation of inhaled ultrafine particles to the brain. *Inhalation Toxicol* 16: 437–445, 2004.

27. G Oberdörster. Pulmonary effects of inhaled ultrafine particles. *Int Arch Occup Environ Health* 74: 1–8, 2000.

28. G Oberdörster. Toxicology of ultrafine particles: *in vivo* studies. *Phil Trans R Soc Lond Ser A Math Phys Eng Sci* 358: 2719–2740, 2000.

29. K Donaldson, V Stone, A Clouter, L Renwick, and W MacNee. Ultrafine particles. *J Occup Environ Med* 58: 211–216, 2001.

30. K Donaldson, V Stone, PS Gilmour, DM Brown, and WNE MacNee. Ultrafine particles: mechanisms of lung injury. *Phil Trans R Soc Lond Ser A Math Phys Eng Sci* 358: 2741–2748, 2000.

31. K Donaldson, D Brown, A Clouter, R Duffin, W MacNee, L Renwick, L Tran, and V Stone. The pulmonary toxicology of ultrafine particles. *J Aerosol Med* 15: 213–220, 2002.

32. PJA Borm. Particle toxicology: from coal mining to nanotechnology. *Inhalation Toxicol* 14: 311–324, 2002.

33. SA Murphy, KA Berube, and RJ Richards. Bioreactivity of carbon black and diesel exhaust particles to primary Clara and type II epithelial cell cultures. *Occup Environ Health Med* 56: 813–819, 1999.

34. PricewaterhouseCoopers, Thomson Venture Economics, National Venture Capital Association Money Trade Survey, *Smalltimes*. From D Forman, Nanotech rides a rising tide. *Smalltimes* 4: 18–21, 2004.

35. Center for Responsive Politics. *Pharmaceuticals/Health Products: Long-Term Contribution Trends*. Washington, D.C. http://www.opensecrets.org/industries/indus.asp? Ind=H04.

36. Center for Responsive Politics. *Chemical and Related Manufacturing: Long-Term Contribution Trends*. Washington, D.C. http://www.opensecrets.org/industries/ indus.asp?Ind=N13.

37. NanoBusiness Alliance. New York. http://www.nanobusiness.org/.

38. BJ Feder. Nanotechnology group to address safety concerns. *New York Times*, Section C: 6, July 7, 2003.

39. A Hett et al. *Nanotechnology: Small Matter, Many Unknowns*. Swiss Reinsurance Company, Zurich. 2004. http://www.swissre.com/INTERNET/pwswpspr.nsf/fm BookMarkFrameSet?ReadForm&BM=. /vwAllbyIDKeyLu/MSHS-4THQED?Open Document.

40. R Highfield. Prince asks scientists to look into 'grey goo.' *London Daily Telegraph*. June 5, 2003. http://www.telegraph.co.uk/news/main.jhtml?xml=/news/2003/06/05/nano05.xml.

41. Staff and Agencies. Prince sparks row over nanotechnology (commentary). *Guardian*. London. April 28, 2003. http://education.guardian.co.uk/higher/research/story/0,9865,945158,00.html.

42. C Lucas. Women face greater exposure to 'grey goo' science: health and beauty products treat women as guinea pigs. Press release, U.K. Green Party. May 22, 2003. http://www.greenparty.org.uk/index.php?nav=news&n=568.

43. C Lucas. We must not be blinded by science. *Guardian*. London. June 12, 2003. http://www.guardian.co.uk/comment/story/0,3604,975427,00.html.

44. ETC Group. No Small Matter II: The Case for a Global Moratorium. Winnipeg, Canada. 2003. http://www.etcgroup.org/documents/Comm_NanoMat_July02.pdf.

45. ETC Group. The Big Down: Atomtech: Technologies Converging at the Nano-Scale. Winnipeg, Canada. 2003. http://www.etcgroup.org/documents/TheBigDown.pdf.

46. AH Arnall. Future technology, today's choices: nanotechnology, artificial intelligence and robotics; a technical, political, and institutional map of emerging technologies. Greenpeace Environmental Trust. London. 2003.

47. GeneWatch U.K. Comments to the Royal Society and Royal Academy of Engineering Working Group on Nanotechnology. July 2003. http://www.nanotec.org.uk/evidence/57aGenewatch.htm.

48. National Center for Environmental Research. Impact of manufactured materials on human health and the environment: Science to Achieve Results (STAR) Program. U.S. Environmental Protection Agency. http://es.epa.gov/ncer/rfa/current/2003_nano.html.

49. National Institute of Environmental Health, National Institutes of Health. Substances Nominated to the NTP for Toxicological Studies and Testing: Recommendations of the NTP Interagency Committee for Chemical Evaluation and Coordination (ICCEC) on June 10, 2003. http://ntp-server.niehs.nih.gov/NomPage/2003 Noms.html.

50. Center for Biological and Environmental Nanotechnology. Rice University, Houston, TX. http://www.ruf.rice.edu/~cben/.

51. VL Colvin. Testimony before U.S. House of Representatives Committee on Science, hearing on Societal Implications of Nanotechnology. 108th Congress, Washington, D.C. April 9, 2003. http://www.ruf.rice.edu/~cben/ColvinTestimony040903.shtml.

52. U.K. Nanotechnology Working Group. The Royal Society and the Royal Academy of Engineering. London. http://www.nanotec.org.uk/workingGroup.htm.

53. Nano-Pathology Project. Quality of Life and Management of Living Resources Programme. European Community. Brussels.

54. Nanoderm Project. Quality of Life and Management of Living Resources Programme. European Community. Brussels.

55. Nanosafe Project. Competitive and Sustainable Growth Programme. European Community. Brussels.

56. *Chemical Abstract Service*. American Chemical Society. Columbus, OH. http://www.cas.org/EO/regsys.html.

57. Federal Hazardous Substance Act. U.S. Consumer Product Safety Commission. http://www.cpsc.gov/businfo/fhsa.html.

58. Toxic Substance Control Act. U.S. Environmental Protection Agency. http://www.epa.gov/region5/defs/html/tsca.htm.

59. Nanotechnology and Regulation: A Case Study Using the Toxic Substance Control Act (TSCA). Foresight and Governance Project, Woodrow Wilson International Center for Scholars, Washington, D.C. 2003. http://www.environmentalfutures.org/nanotsca_final2.pdf.

60. CD Stone. *Where the Law Ends: The Social Control of Corporate Behavior*. New York: Harper Torchbooks. 1975.

61. C Perrow. *Normal Accidents*. New York: Basic Books. 1984.

62. Center for Legal Policy. Trial Lawyers Inc: a Report on the Lawsuit Industry in America. Manhattan Institute for Policy Research. 2003. http://www.triallawyers-inc.com/html/part01.html.

63. GJ Stigler. The theory of economic regulation. *Bell J Econ Mgt Sci* 2: 3–21, 1971.

64. P Berg, D Baltimore, HW Boyer, SN Cohen, RW Davis, DS Hogness, D Nathans, R Roblin, JD Watson, S Weissman, and ND Zinder. Potential biohazards of recombinant DNA molecules, *Science* 185: 303, 1974.

65. Treaty of Maastricht on the European Union. http://europa.eu.int/en/record/mt/title2.html.

66. Commission of the European Communities. Communication on the precautionary principle. Brussels, February 2, 2000. COM (2000) 1. http://europa.eu.int/comm/dgs/health_consumer/library/pub/pub07_en.pdf.

Index

took the lead in defending—and extending—the civil rights of black Americans.

Far from seeking to strengthen or extend racial discrimination, Free Soilers frequently went out of their way to debunk conventional stereotypes of black inferiority and to preach the "equality of all men, of every climate, color, and race." None did so more passionately than Wisconsin's Free Soil congressman, Charles Durkee. "Sir," he lectured the House of Representatives in 1850, "you may take from . . . [a man] his wife, his children, and his friends, and put handcuffs on his wrists and fetters on his feet, and brand him as your property . . . yet he is still a *man*—he is still our brother." In the same spirit, the Washington *National Era* explained that Free Soilers were "opposed to the spirit of caste, whether its elemental idea be a difference of color, birth, or condition—because its inevitable tendency is to create or perpetuate inequality of natural rights."[32]

Although conspicuously reticent with regard to social equality, and often patronizing in their relations with blacks, many Free Soilers perceived a grating disparity between the Declaration of Independence and the Bill of Rights, which spoke of natural rights belonging to *all* men, and the web of discriminatory laws and customs that in every corner of the land consigned black Americans to second-class citizenship. As one Ohio Free Soiler caustically observed, " 'All men by nature are free and independent,' don't look well by the side of 'every White male citizen.' "[33] Accordingly, Free Soilers in many states and on many occasions braved taunts of "amalgamationist" and "woolly-head" and joined out-and-out abolitionists in the fight against institutionalized racism. And since at mid-century something like nineteen of twenty Northern Negroes lived in states that banned or severely restricted their right to vote, suffrage reform was an obvious starting point.

Wisconsin witnessed the most nearly successful campaign for color-blind voting requirements. Though the 1850 census showed only 635 blacks, scattered widely about the state, the question of their enfranchisement inevitably surfaced during debates over the state's first constitution. Most Democrats followed the lead of Moses M. Strong, who declared himself "teetotally opposed" to any form of black suffrage. A great many Whigs, however, as well as most Liberty men and Free Soilers, endorsed equal suffrage from the outset. At its annual meeting in 1848, the Wisconsin Liberty Association (virtually all of whose members soon merged with the Free Soil party) resolved that suffrage was an "inalienable right" belonging to all men, that to deny this right to any would be to open the door to disfranchisement for all, and that the principle that had robbed the Negro of one right "only needs scope and opportunity to rob him of *all* rights." Early the next year Wisconsin Free Soilers boldly proclaimed: "We are in favor of equal and impartial suffrage, and are the friends of man and the advocates of human rights the world over." The issue was "one of vital importance," the *Wisconsin Free Democrat* contended. "Practically," the paper admitted, "it